A Field Guide for the Identification of Invasive Plants in Southern Forests

James H. Miller, Erwin B. Chambliss,
and Nancy J. Loewenstein

Acknowledgments

The clarity and accuracy of descriptions and nomenclature have been measurably improved by the sizable and appreciated reviews of the following experts. Reviewers of this edition: Curtis Hansen, Auburn University, Department of Biological Sciences, Auburn, AL; Ron Lance, plant consultant, Asheville, NC; Steven T. Manning, Invasive Plant Control, Inc., Nashville, TN; Carey Minteer, formerly with the University of Georgia's Center for Invasive Species and Ecosystem Health, Tifton, GA; and Fred Nation, Weeks Bay National Estuarine Research Reserve, Fairhope, AL.

Reviewers of prior edition: Kristine Johnson, Great Smoky Mountains National Park, Gatlingburg, TN; Fred Nation, Weeks Bay National Estuarine Research Reserve, Fairhope, AL; Johnny Randall, North Carolina Botanical Gardens, Chapel Hill, NC; Jack Ranney, formerly with the University of Tennessee, Knoxville, TN.

All Plant Images by James H. Miller and Posted on Bugwood.org with Crucial Contributions by:

Ted Bodner, copyrighted by the University of Georgia Press for "Forest Plants of the Southeast and Their Wildlife Uses" and used by permission, Champaign, IL

Patrick Breen, Oregon State University, Department of Horticulture, Corvallis, OR, Bugwood.org

Steven J. Baskauf, Vanderbilt University, Nashville, TN

Charles T. Bryson, USDA Agricultural Research Service, Southern Weed Science Research Unit, Stoneville, MS, Bugwood.org

John Cardina, The Ohio State University, Columbus, OH, Bugwood.org

Gerald D. Carr, Carr Botanical Consultation, Bugwood.org

Erwin Chambliss, U.S. Forest Service, Southern Research Station, Auburn, AL

Steve Dewey, Utah State University Extension, Logan, UT, Bugwood.org

Thomas Ellis, Jr., Baldwin County Forestry Planning Committee, Bay Minette, AL

Stephen F. Enloe, Department of Agronomy & Soils, Auburn University, Auburn, AL

Chris Evans, River to River Cooperative Weed Management Area, Marion, IL, Bugwood.org

John Everest, Department of Agronomy & Soils, Auburn University, Auburn, AL

Nancy Fraley, U.S. Department of the Interior, National Park Service, Asheville, NC

Great Smoky Mountains National Park Resource Management Archive, USDI National Park Service, Gatlinburg TN, Bugwood.org

Michael Jordan, Alabama Department of Environmental Management, Montgomery, AL

Steven Katovich, U.S. Forest Service, Forest Health Protection, Bugwood.org

Ron Lance, plant consultant, Asheville, NC

Nancy Loewenstein, School of Forestry and Wildlife Sciences, Auburn University, Auburn, AL, Bugwood.org

Steven T. Manning, Invasive Plant Control Inc., Nashville, TN, Bugwood.org

Leslie J. Mehrhoff, University of Connecticut, Bugwood.org

David J. Moorhead, Center for Invasive Species and Ecosystem Health, University of Georgia, Bugwood.org

Fred Nation, Weeks Bay National Estuarine Research Reserve, Fairhope, AL, Bugwood.org

Ohio State Weed Lab Archive, The Ohio State University, Bugwood.org

Richard Old, XID Services, Incorporated, Bugwood.org

Christopher Oswalt, U.S. Forest Service, Southern Research Station, Forest Inventory and Analysis, Knoxville, TN

Pennsylvania Department of Conservation and Natural Resources - Forestry Archive, Harrisburg, PA, Bugwood.org

Corrie Pieterson, University of Florida, School of Forest Resources and Conservation, Gainesville, FL

John M. Randall, The Nature Conservancy, University of California, Davis, CA, Bugwood.org

Barry Rice, sarracenia.com, Bugwood.org

Amy Richard, University of Florida, Bugwood.org

Jan Samanek, State Phytosanitary Administration, Bugwood.org

Michael Shephard, U.S. Forest Service, Bugwood.org

John Schwegman, Illinois Department of Natural Resources, Springfield, IL

Forest and Kim Starr, U.S. Geological Survey, Bugwood.org

Dan Tenaglia, Missouriplants.com, Bugwood.org

Robert Vidéki, Doronicum Kft., Bugwood.org

Warner Park Nature Center, Metropolitan Board of Parks and Recreation, Nashville, TN

Gil Wojciech, Polish Forest Research Institute, Bugwood.org

Plant Names from:

USDA Natural Resources Conservation Service's Plants Database: http://plants.usda.gov with recent published modifications.

For sale by the Superintendent of Documents, U.S. Government Printing Office
Internet: bookstore.gpo.gov Phone: toll free (866) 512-1800; DC area (202) 512-1800
Fax: (202) 512-2104 Mail: Stop IDCC, Washington, DC 20402-0001

ISBN 978-0-16-085733-1

Contents

Explanation of Codes and Conventions Used in Species Descriptions

Scientific name ↙ ***International code*** ↙ ***Family name*** ↙ ***Common name*** ↙ ***Forest Inventory and Analysis survey code*** ↘

Firmiana simplex (L.) W. Wight **FISI2** **Sterculiaceae** **Chinese Parasoltree** **0998**

Synonym: *F. platanifolia* (L. f.) Schott & Endl., *Sterculia platanifolia* L. f.
Other common names: phoenix tree, varnish-tree

Plant. Deciduous upright tree to 50 feet (16 m) in height and 2 feet (60 cm) in trunk diameter with stout alternate branches. Leaves dark green, large, 3- to 5-lobed resembling a maple leaf, and petioles almost as long as the leaf. Terminal clusters of tan and yellow flowers appear in midsummer to quickly yield unusual pods that split into 4 petal-like sections with pea-sized seeds attached to the upper margins. Leaves turn yellow in fall and branched showy fruit stalks remain during winter.

Stem. Twigs and branches stout, glossy green when rapidly growing or dull green becoming grayish green from a wax coating. Buds large and round with numerous

Chinese wisteria, Chinese privet, Chinese lespedeza, and Japanese honeysuckle replace forest stands.

Tallowtree, silktree, Chinese privet, Japanese honeysuckle, and cogongrass can comprise a new forest stand type.

Introduction

Invasions of nonnative plants into southern forests continue to go largely unchecked and only partially monitored. Small forest openings, forest road right-of-ways, and areas under and beside forest canopies are often occupied by invasive nonnative plants. These infestations increasingly erode forest productivity, hindering forest use and management activities, degrading diversity and wildlife habitat. Often called nonnative, exotic, nonindigenous, alien, or noxious weeds, nonnative invasive plants occur as trees, shrubs, vines, grasses, ferns, and forbs. Some have been introduced into this country accidentally, but most were brought here as ornamentals or for livestock forage. These robust plants arrived without their natural predators of insects, diseases, and animals that tend to keep native plants in natural balance. Many have hybridized and undergone plant breeding to become more aggressive, predator resistant and resilient, drought tolerant, and cold hardy. Now, they increase across the landscape with little opposition beyond the control and reclamation measures applied by landowners, managers, and agencies on individual land holdings. An increased awareness of the threat has resulted in growing networks of concerned individuals, agencies, governments, and companies aimed at stopping plant invasions across landscapes and restoring formerly infested lands.

The objective of this book is to provide information on accurate identification of 56 plants or groups (like the many invasive roses) that are aggressively invading forests of the 13 Southern States at alarming rates. It also lists other nonnative invasive plants that are of growing concern. A companion booklet, "A Management Guide for Invasive Plants of Southern Forests," explains control recommendations, selective application procedures, and prevention measures for these and other plant invaders. The text and photographs for 33 species were originally developed for the 2003 edition to assist in the first regionwide survey and monitoring of these invading species. Survey work was conducted by the Forest Service, U.S. Department of Agriculture, Forest Inventory and Analysis Research Work Unit of the Southern Research Station in collaboration with State forest management agencies. The four-number survey codes are given for these species as well as the international plant codes for all species when available (see opposite page). The latest data for this ongoing survey can be viewed and downloaded at http://srsfia2.fs.fed.us/data_center/index.shtml.

Accurate identification of nonnative invasive plants is critical for recognition of new entries and positive control of existing infestations. Many invasive species have native and nonnative plants that closely resemble them and most have been commercially bred to yield many varieties with varying traits. A section is included for each invasive species giving traits to distinguish them from their "look-a-likes." Use these traits to safeguard valuable native plants and for directing controls toward the appropriate invasive plant species.

Brazilian Peppertree

Steven T. Manning

Steven T. Manning

Michael Jordan

Schinus terebinthifolius Raddi **SCTE** **Anacardiaceae**
Other common names: Brazilian pepper, Florida holly, Christmas berry

Brazilian Peppertree **FL28**

Plant. Evergreen shrub or small tree to 40 feet (12 m) in height often in dense infestations with many short trunks or arching stems of contorted branches. Drooping, odd-pinnately compound leaves that smell of turpentine when crushed. Female plants have many multi-branched clusters of small, whitish flowers in summer and fall that yield clusters of spherical, red and pepper-smelling fruit in winter. **Caution: All parts of the plant can cause skin rash or airway irritation in sensitive people.**

Stem. Twigs and root sprouts yellow green with V-shaped leaf scars, becoming smooth gray-brownish branches that become entangled and tend to droop. Older bark braided with grayish ridges and reddish-brown fissures.

Leaves. Evergreen and thick, alternate, 3 to 12 inches (8 to 30 cm) long having reddish, sometimes winged midribs, odd-pinnately compound with 3 to 13 sessile, ovate to elliptic leaflets, finely toothed, each 1 to 2 inches (2.5 to 5 cm) long, shiny and green above with light-green midveins and lateral veins and blade paler beneath. Often drooping and emit an aroma of pepper or turpentine when crushed.

Flowers. September to November and other times. Axillary and terminal on new growth. Multi-branched clusters of many tiny flowers with 5 white petals and yellow centers. Male and female flowers on separate plants (dioecious).

Fruit and seeds. December to August. Berrylike, spherical drupes, 0.2 to 0.3 inch (6 to 8 mm) wide, in profuse clusters, light green ripening quickly to bright red and then dark red, fleshy and juicy becoming papery, containing 1 dark brown 0.1- inch (0.3 mm) long seed.

Ecology. Forms dense thickets, spreading by many root sprouts that yield entangled stems and branches with abundant foliage that contains allelopathic chemicals to exclude other plants and animals. Burns hot. Tolerant to a wide variety of growing conditions, but grows best in moist soils. Persists in shade with rapid growth in full sun. Producing seed as early as 3 years. Abundant seed are spread by birds, with seedlings able to establish in shade. Presently limited by cold, but spreading northward with warming trends.

Resembles the nonnative peppertree (*S. molle* L.) escaped in FL, TX, and CA, but its 19 to 40 leaflets are narrowly lance-shaped. **Also might resemble** another rash causing shrub, poison sumac [*Toxicodendron vernix* (L.) Kuntze] that frequents similar wet habitats and has jutting, odd-pinnately compound leaves with sharp-tipped leaflets and reddish petioles and stalks.

History and use. Native to South America and introduced in the 1840s to FL as an ornamental and widely sold and planted until recently.

Distribution. Extensive and dense infestations in FL and south TX.

June

Forest and Kim Starr

Callery Pear, Bradford Pear

Pyrus calleryana Decne. **PYC80** **Rosaceae**

Callery Pear, Bradford Pear

Plant. Deciduous tree to 60 feet (18 m) in height and boles 2 feet (0.6 m) in diameter, widely planted as an ornamental tree due to its uniform teardrop crown shape, early spring cover of white blossoms, and brilliant orange-to-red fall foliage. Leaves ovate and long-petioled, alternate and tufted on stubby thorned or nonthorned branchlets. Abundant (rarely few) small pears in fall and winter spread by birds and other animals, with varying seed fertility. Several commercial sterile varieties can cross-pollinate with certain other pear species in close proximity to produce fertile hybrids. Also, fertile varieties are commonly used as a rootstock for grafting most pear species and can dominate after grafted crowns are damaged. Major branches fork from trunk at narrow angles, often splitting at the juncture after wind and ice storms.

Stem. Twigs initially hairy, stubby, and tipped by a sharp thorn in escaped plants, being longer branched with few thorns in cultured, planted varieties. Terminal and lateral bud scales loose, gray-hairy and elongated to 0.5 inch (1.2 cm). Twigs reddish brown to gray with scattered or many light dots (lenticels) that elongate with age to form horizontal light lines on branches and boles. Bark of trunk increasingly vertically fissured, becoming cross-checked with age. Stems and boles often dappled with gray and green lichens.

Leaves. Alternate, often tufted on short branchlets. Initially circular and hairy, maturing to glossy and ovate or slightly cordate with a tapering tip, 1.5 to 3.5 inches (4 to 9 cm) long and wide, leathery with finely crenate and wavy margins sometimes having a pronounced tip. Dark green above and light green below, becoming brilliantly red, yellow, to maroon in fall. Thin petioles 1 to 2 inches (2.5 to 5 cm) long with thin stipules that shed early.

Flowers. March to May. Large clusters of brilliantly white, 5-petaled flowers, 1 inch (2.5 cm) across with many jutting, maroon-tipped anthers, appearing before leaves, tufted often from midthorn, covering trees to make conspicuous invaders in early spring. Emit an unpleasant odor and can cause allergic sinus reactions.

Fruit and seeds. June to February. Persistent clusters of small pears (pomes), 0.3 to 0.5 inch (8 to 12 mm) long and wide, green ripening to tan then maroon with numerous surface speckles, dangling on thin, 1-inch (2.5 cm) long stems. Fleshy, tart but edible, containing 2 to 6 maroon seeds and numerous stone cells.

Ecology. Rapid growing, forms thickets and dense stands by root sprouts. Tolerates partial shade and a variety of soils from wet to droughty. Hybridizes with other callery pear varieties and Asian pear (*P. betulifolia* Bunge). Pollinated by insects. Fertility and fruit production vary widely, but close colonies result in more viable fruit. Fruiting starting at 3 years. Spreads by prolific bird- and animal-dispersed seeds. Seeds require 60 to 90 days of cold to germinate, which can occur in the fruit. Widely invading over a large range due to cultivars that have been bred for cold hardiness.

Resembles leaves of both common pear (*Pyrus* spp.) that has larger fruit and Chinese tallowtree [*Triadica sebifera* (L.) Small] that has dangling spring flowers and clusters of white fruit in fall and winter. **Also resembles** flowers of another invasive, common pearlbush [*Exochorda racemosa* (Lindl.) Rehder] that blooms in early spring with similar bright white flowers that occur in racemes and leaves are elliptic and serrated above the middle.

History and use. Native to China, Korea, and Taiwan, the original "Bradford" cultivar was introduced into Maryland in 1908 for breeding resistance to fire blight disease into fruiting pears, although not successful. Since the 1950s about 20 ornamental cultivars have been developed and are still widely sold and planted. All capable of producing viable seeds.

Distribution. Found as single trees and some dense infestations throughout the region.

Camphortree

December

June

June

June

June

October

November

November

Cinnamomum camphora (L.) J. Presl **CICA** **Lauraceae**

Other common names: camphor laurel, shiu leaf, gum camphor, laurel camphor

Plant. Evergreen tree 60 to 100 feet (18 to 30 m) in height and boles up to 5 feet (1.5 m) in diameter, with a round spreading crown formed by large branches radiating from midtree. Also found in dense thickets from root sprouts and seedlings. Glossy, lanceolate leaves and slender, green to reddish-brown twigs having a camphor odor when crushed, cut, or bruised. Abundant clusters of spherical, black drupes in fall to winter.

Stem. Twigs slender, smooth, and green to yellowish green sometimes tinged with red, with scattered light dots (lenticels). Corky ridges encircle junctures of new growth with jutting leaf scars on prior year's growth. Buds 0.5 inch (1.2 cm) long, sharp-pointed, with overlapping scales. Branches green eventually reddish, smooth and glossy, increasingly covered with gray corky ridges and plates that fissure with age. Bark light grayish brown and widely fissured, developing vertical plates with age.

Leaves. Alternate but more clustered at twig tips, evergreen, leathery and glossy with a camphor odor when crushed, lanceolate, ovate to obovate, 2 to 4 inches (5 to 10 cm) long and 0.8 to 2 inches (2 to 5 cm) wide. Burgundy when young turning dark green with light-green veins above, whitish green beneath with tiny, brownish glands in axils of main veins, also evident as light dots on the upper surface. Margins entire, yellowish green, and wavy. Petioles slender to 1 inch (2.5 cm) long.

Flowers. May. Slender, stalked, axillary panicles, 2 to 3 inches (5 to 7 cm) long with perfect tiny, greenish-white, fragrant flowers, with both male and female parts.

Fruit and seeds. July to February. Many fleshy, rounded drupes, 0.3 inch (0.7 cm) wide, dangling, green turning black with a funnel-shaped, green base.

Ecology. Rapid growing, forming thickets and dense stands in fencerows, disturbed sites, and adjacent upland forests. Grows in well-drained, acid to basic soils and can withstand partial shade. Cannot tolerate extended cold temperatures below 20 °F (-6 °C). Colonizes by root sprouts and spreads by prolific animal- and water-dispersed seeds.

Resembles another nonnative invasive, Chinese tallowtree [*Triadica sebifera* (L.) Small], which is deciduous and emits no camphor odor. **Also resembles** Carolina laurelcherry [*Prunus caroliniana* (Mill.) Ait.], considered an invasive native evergreen tree that has smooth, dark gray bark and leaves that emit an almond scent when crushed.

History and use. Introduced in 1727 from Japan, China, and Taiwan as an ornamental and herbal plant, although it is toxic to humans in large doses. The wood widely used in Asia for chests, panels, and lumber.

Distribution. Found as scattered plants or in dense infestations in southeast TX; south MS, AL, and GA; east SC and NC; and central and north FL.

Camphortree

0855

May

Chinaberrytree

May

July

November

January

July

July

November

July

Melia azedarach L. **MEAZ** **Meliaceae**
Other common names: Persian lilac, pride-of-India, chinaberry

Chinaberrytree **0993**

Plant. Deciduous tree to 50 feet (15 m) in height and 2 feet (60 cm) in diameter, much branched with multiple boles, lacy dark green leaves having a musky odor, and clusters of lavender flowers in spring yielding persistent, yellow berries. **Caution: Fruit (berries) is poisonous.**

Stem. Twigs stout, glossy greenish brown with light dots (lenticels). No terminal bud. Numerous broad, V-shaped, raised leaf scars with 3 bundle scars below a domed fuzzy bud. Bark dark brown and becoming increasingly fissured with age. Wood soft and white.

Leaves. Alternate spiraled, bipinnately compound, 1 to 2 feet (30 to 60 cm) long and 9 to 16 inches (23 to 40 cm) wide. Leafstalk lime green with base slightly clasping stem. Each leaflet lanceolate with tapering tips, 1 to 3 inches (2.5 to 8 cm) long and 0.5 to 1.2 inches (1.2 to 3 cm) wide. Margins varying from entire to coarsely crenate to serrate and wavy. Glossy dark green with light-green midvein above and pale green with lighter green midvein beneath, becoming golden yellow in fall.

Flowers. March to May. Showy panicles from lower axils of new stems. Five pinkish-lavender to whitish petals, stamens united in dark-purple tube. Five green sepals. Fragrant.

Fruit and seeds. July to January. Berrylike spherical drupe 0.5 to 0.7 inch (1.2 to 1.8 cm) wide persisting through winter and containing a stone with 1 to 6 seeds. Light green turning yellowish green then yellowish tan. Poisonous to humans and livestock.

Ecology. Common on roadsides, fencerows, at forest margins, and around old homesites, but rare at high elevations. Somewhat shade and flood tolerant. Occurs from uplands to marshes. Viable seed can be produced by 4- and 5-year-old plants. Forms colonies from root sprouts, sprouts from root collars, and seedlings. Spreads by abundant bird-dispersed seeds. Germination suppressed under mature trees but seeds germinate following parent tree removal.

Resembles common elderberry [*Sambucus nigra* L. ssp. *canadensis* (L.) R. Bolli], a spreading crowned shrub with once pinnately compound leaves having leaflets with finely serrate margins and green to dark-purple berries in flat-topped clusters.

History and use. Introduced in the mid-1800s from Asia. Widely planted as a traditional ornamental around homesites. Extracts potentially useful for natural pesticides. Though softer and less durable than many other members of the mahogany family, chinaberry wood has an attractive grain and takes a good finish for inexpensive cabinetry.

Distribution. Found throughout the region with the most frequent forest infestations in east TX; west LA; south to central MS, AL, GA, and SC; and north FL.

June
November

Chinese Parasoltree

Firmiana simplex (L.) W. Wight **FISI2** **Sterculiaceae**

Synonym: *F. platanifolia* (L. f.) Schott & Endl., *Sterculia platanifolia* L. f.
Other common names: phoenix tree, varnish-tree

Plant. Deciduous upright tree to 50 feet (16 m) in height and 2 feet (60 cm) in trunk diameter with stout alternate branches. Leaves dark green, large, 3- to 5-lobed resembling a maple leaf, and petioles almost as long as the leaf. Terminal clusters of tan and yellow flowers appear in midsummer to quickly yield unusual pods that split into 4 petal-like sections with pea-sized seeds attached to the upper margins. Leaves turn yellow in fall and branched showy fruit stalks remain during winter.

Stem. Twigs and branches stout, glossy green when rapidly growing or dull green becoming grayish green from a wax coating. Buds large and round with numerous overlapping fuzzy scales, initially reddish turning maroon in late winter. Large leaf scars, raised rimmed and circular to oblong, topped by small linear stipule scars on both sides and a fuzzy rounded bud in winter. Pronounced rings of raised bud scars encircle twigs. Pith white, spongy, and continuous. Bark grayish tan, tight, with vertical green to orange shallow stripes in late winter becoming roughened with age. Pronounced eye-shaped branch scars occur along the trunk.

Leaves. Alternate, 8 to 12 inches (20 to 30 cm) long and somewhat wider and larger on vigorous young shoots, 3 to 5 shallow to deep lobes and tips, and a cordate base. Dark green with light-green palmate veins above a whitish patch where they join with the petiole and softly hairy beneath. Margins entire. Petioles slightly rough, 12 to 20 inches (20 to 50 cm) long. Leaves turning yellow in the fall.

Flowers. May to July. Large, branched panicles over 2 feet (60 cm) across of tan and yellow flowers, from the base of new growth. Branches green to pale green. Separate male and female flowers occur in each cluster and open at varying times. Slightly fragrant. Male flowers turn reddish pink before petals fall.

Fruit and seeds. June to April. Quickly after flowering, 1 to 5 pea-size dry fruit appear along the upper margins of unique drooping petal-like curved sections, 2.5 to 3 inches (6 to 8 cm) long and 4 to 5 in a flower-like group, initially light green turning tan, thin and wafery. Panicle branches with fruit drop throughout winter leaving star-shaped woody flower bases at branch ends.

Ecology. Occasional ornamental plantings in the southern half of the region are a source of escaped plants in surrounding roadsides, riparian areas, and forest margins. Rapid early growth. Readily self-pollinates and self-seeds, and thus it is thought that a single, reproductive age tree can produce an entire colony. Nectaries in flowers suggest insect pollination. Spreads and forms infestations by wind- and water-dispersed seeds. The high fat seed has not been observed as spread by wildlife. Root sprouts have not been observed.

Resembles tungoil tree [*Vernicia fordii* (Hemsl.) Airy-Shaw], which has similar shaped leaves but has pairs of conspicuous dark glands where the petiole joins the leaf blade and white milky sap.

History and use. Introduced in 1757 from China. Planted as an ornamental. The wood is used for the soundboards of several Chinese instruments.

Distribution. Found as escaped plants and small groups more frequently in the coastal States but throughout the region except in KY and north TN, with northward spread expected.

Chinese Parasoltree **0998**

June

Glossy Buckthorn

Leslie J. Mehrhoff

Richard Old

Gil Wojciech

Steven T. Manning

Frangula alnus Mill. FRAL4 ***Rhamnaceae***

Synonym: *Rhamnus frangula* L.
Other common names: alder buckthorn, glossy false buckthorn, columnar buckthorn, fen buckthorn

Plant. Deciduous shrub with many sprouts from the base, or small tree, 6 to 24 feet (12 m) in height, up to 10 inches (25 cm) in diameter with glossy bark (thus the common name), thornless stems and an oval, much-branched crown. Alternate, oval leaves appear early in spring and linger green into fall, dark green and glossy above with distinct parallel lateral veins. Stemmed clusters of tiny white flowers in summer yield spherical, green berrylike fruit that turn red then black in the fall.

Stem. Twigs alternate, initially reddish, and slightly gray fuzzy with scattered white dots (lenticels), becoming hairless and gray to brownish gray with light dots in lengthwise bands that become raised and eventually turn into lengthwise, shallow fissures on larger stems. Buds pointed and fuzzy. Leaf scars raised with 3 bundle scars. Cut twigs have yellow sapwood and pinkish to orange centers.

Leaves. Alternate, tardily deciduous, narrowly elliptic to oblong to obovate, 2 to 4 inches (5 to 10 cm) long and one-half as wide with 8 to 9 pairs of parallel lateral veins curving upward at the ends to follow the margin. Margins entire and somewhat wavy. Shiny green above, paler and slightly hairy beneath, turning greenish yellow in the fall. Petioles reddish and hairy, 0.25 to 0.5 inch (6 to 12 mm) long.

Flowers. May to September. Axillary, stemmed clusters of tiny flowers with 5 whitish petals barely jutting from a bell-shaped green calyx, appearing on new growth after the leaves.

Fruit and seeds. May to November. Berrylike, spherical drupes, 0.3 to 0.4 inch (8 to 12 mm) wide, light green ripening to red then black in late summer, fleshy, containing 2 to 3 ovoid seeds.

Ecology. Tolerant of a wide variety of growing conditions from wet to dry and basic to acidic soils. Persists in shade with rapid growth in full sun to produce seed as early as year 3. Wide spreading with many sprouts, leafing out early in spring and retaining foliage late in fall leading to the exclusion of other forest plants. Abundant seed spread by birds, with seedlings able to establish in shade. Invades forest edges and understories.

Resembles the native Carolina buckthorn [*Frangula caroliniana* (Walt.) Gray], but its leaf margins are finely serrated (serrulate) and leaves are 3 times as long as wide. **Also resembles** common buckthorn (*Rhamnus cathartica* L.), another invasive that has finely serrated leaf margins but with 3 to 5 pairs of curved lateral veins, thorn-tipped twigs, only 4 petals, and 3 to 4 seeds. **Also resembles** the native alderleaf buckthorn (*R. alnifolia* L'Hér.), a shrub to 6 feet (2 m) tall and wide with crenate-serrate margins, 6 to 9 paired lateral veins, and hairless beneath.

History and use. Initially introduced from Eurasia and North Africa in the mid-1800s as an ornamental.

Distribution. Found as scattered plants or dense infestations in VA, TN, and NC.

Glossy Buckthorn

Paper Mulberry

November

July

June

April

Gerald D. Carr

July

Amy Richard

June

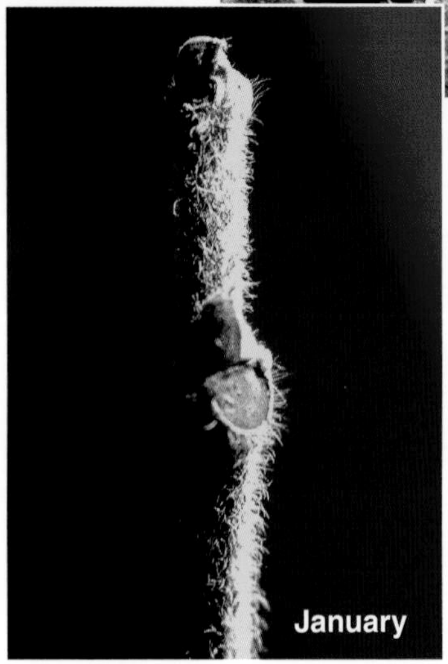

January

April

September

Steven J. Baskauf

Broussonetia papyrifera (L.) L'Hér. ex Vent. **BRPA4** **Moraceae**
Synonym: *Morus papyrifera* L.

Plant. Deciduous, large shrub or tree to 50 feet (15 m) in height and boles to 2 feet (0.6 m) in diameter with a round crown. Often appears as a shrub, forming thickets from root sprouts. Broad oval leaves, sometimes deeply lobed on rapidly growing stems, softly hairy on the lower surface and scruffy above. Downy gray twigs with scattered prominent orange dots (lenticels). Shallow rooted and prone to windthrow. Separate male and female plants (dioecious).

Stem. Twigs moderately stout, zigzagging, dark to light gray, sometimes greenish to reddish brown, covered with silvery down with distinct orange dots (lenticels) when young. Oval leaf scars have protruding rims with stipular scars of light lines on both sides and a hairy, domed flower bud in winter. Branches smooth and mottled gray (green with algae at times) increasingly with protruding orange-tan lenticels, leaf and stipular scars. Pith white with woody diaphragms at nodes. Bark light gray, braided with pale orange to light tan stripes, becoming yellowish with age.

Leaves. Alternate, sometimes opposite or whorled, oval to heart shaped, 3 to 10 inches (7.5 to 25 cm) long, sometimes with 1 to 6 deeply rounded sinuses. Dark green and sandpapery above, whitish and velvety below. Margins finely serrate to sharply toothed except near the base. Petioles 2 to 5 inches (5.0 to 12.5 cm) long, light green, and hairy. Stipules quickly shed.

Flowers. April to May. Male and female flowers appear on separate plants. Clusters of tiny male flowers are elongate, woolly, and drooping, 2.5 to 3 inches (6 to 8 cm) long. Female flowers globular, 1 inch (2.5 cm) wide. Both pale green turning purple.

Fruit and seeds. July to August on female plants (rare in Southeast). Globular and compound, orange turning reddish purple, 0.7 to 1 inch (1.8 to 2.5 cm) wide, with many embedded or protruding tiny, red seeds.

Ecology. Rapid growing, forming thickets and dense stands in fencerows, disturbed sites, and forest edges. Colonizes by root sprouts and spreads by rarely produced animal-dispersed seeds.

Resembles both the native red mulberry (*Morus rubra* L.) and the nonnative white mulberry (*M. alba* L.), which also have a mixture of lobed and unlobed shaped leaves but are not velvety hairy beneath and do not have rounded fruit (theirs are elongated).

History and use. Introduced in the mid-1700s from Japan and China as a rapidly growing shade tree. Used in ancient times by the Chinese to produce fibrous paper (thus the common name "paper").

Distribution. Found as scattered plants or scattered dense infestations throughout the region.

Paper Mulberry

September

Princesstree, Paulownia

Paulownia tomentosa **PATO2** **Paulowniaceae**
(Thunb.) Sieb. & Zucc. ex Steud.
Synonym: *P. imperialis* Siebold & Zucc.
Other common names: royal empresstree, royal paulownia

Plant. Deciduous tree to 60 feet (18 m) in height and 2 feet (60 cm) in diameter with large heart-shaped leaves, fuzzy hairy on both sides, showy pale-violet flowers in early spring before leaves, and persistent pecan-shaped capsules in terminal clusters in summer to winter. Abundant flower buds present on erect stalks over winter.

Stem. Twigs and branches stout, glossy gray brown and speckled with numerous white dots (lenticels). No terminal bud. Lateral leaf scars raised, circular, and becoming larger, dark, and sunken. Bark light-to-dark gray, roughened, and becoming slightly fissured. Stem pith chambered or hollow and wood white.

Leaves. Opposite, heart shaped and fuzzy hairy on both surfaces, 6 to 12 inches (15 to 30 cm) long and 5 to 9 inches (13 to 23 cm) wide. Leaves larger on resprouts, 16 to 20 inches (40 to 50 cm) across, with extra tips often extending from the end of veins. Petioles rough hairy, 2 to 8 inches (5 to 20 cm) long.

Flowers. April to May. Covered with showy erect panicles of pale-violet flowers before leaves in early spring, tubular with 5 unequal lobes. Fragrant. Flower buds fuzzy, linear, becoming ovoid in summer, and persistent on erect stalks over winter.

Fruit and seeds. June to April. Terminal clusters of pecan-shaped capsules 1 to 2 inches (2.5 to 5 cm) long and 0.6 to 1 inch (1.5 to 2.5 cm) wide. Pale green in summer turning tan and eventually black in winter and persistent into spring. Capsules split in half during late winter, each releasing 1,000s of tiny winged seeds.

Ecology. Common around old homes, roadsides, riparian areas, and forest margins in infested areas. Infrequently planted in plantations. One tree can produce 21 million seeds per year with 2,000 per capsule. Viable seed can be produced by 5- and 7-year-old plants. Spreads by wind- and water-dispersed seeds. In the mountains, seed can be dispersed up to 2 miles (3 km). Invades after fire, harvesting, and other disturbances. Forms colonies from root sprouts.

Resembles southern catalpa (*Catalpa bignonioides* Walt.) and northern catalpa [*C. speciosa* (Warder) Warder *ex* Engelm.], which have leaves with sparsely hairy upper surfaces and rough hairy lower surfaces and long slender, persistent fruit. **Also resembles** the invasive paper mulberry [*Broussonetia papyrifera* (L.) L'Hér. ex Vent.] with somewhat smaller leaves, sandpapery feeling above and white velvety hairy below.

History and use. Introduced in the early 1800s from East Asia. Has been widely planted as an ornamental and grown in scattered plantations for speculative high-value wood exports to Japan.

Distribution. Found as escaped plants from ornamental plantings throughout the region and as scattered dense infestations along highways and roadsides to move into disturbed forests in MS, TN, KY, and NC, as well as north AL and north GA.

Princesstree, Paulownia **0712**

April

Russian Olive

Spring

Patrick Breen

Summer

John M. Randall

Summer

Patrick Breen

Winter

Summer

Patrick Breen

Winter

Patrick Breen

Winter

Patrick Breen

Elaeagnus angustifolia L. **ELAN** **Elaeagnaceae**
Other common name: oleaster

Russian Olive

0997

Plant. Deciduous, thorny tree or shrub to 35 feet (10 m) in height with single or multiple boles, many long narrow leaves, and many yellow fruit covered with minute silvery scales. Rare at present in the South while a widespread invasive elsewhere in the United States. Most often confused with autumn olive (*E. umbellata* Thunb.).

Stem. Twigs slender, thorny, and densely silver scaly in the first year becoming glossy and greenish. Branches smooth and reddish brown. Pith pale brown to orange brown. Bark dark brown and densely fissured.

Leaves. Alternate, long lanceolate to oblanceolate measuring 1.5 to 4 inches (4 to 10 cm) long and 0.4 to 1.2 inches (1 to 3 cm) wide. Margins entire (rarely toothed). Green to slightly silvery above with dense silver scales beneath. Petioles short and silvery.

Flowers. April to July. Axillary clusters, each with 5 to 10 silvery-white to yellow flowers. Tubular with 4 lobes. Fragrant.

Fruit and seeds. August to October. Drupelike, hard fleshy fruit 0.5 inch (1.2 cm) long, resembling an olive. Light green to yellow with silvery scales. One nutlet in each fruit.

Ecology. Found as rare plants in city forests, disturbed areas near forests, and escapes from surface mine plantings. Thrives in sandy floodplains. Shade intolerant. Spreads by bird- and animal-dispersed seeds. A nonleguminous nitrogen fixer.

Resembles silverthorn or thorny olive (*E. pungens* Thunb.), which is an evergreen with brown scaly and hairy twigs, flowers in late fall producing reddish silver-scaly drupes in spring. **Also resembles** autumn olive (*E. umbellata* Thunb.), a widespread invasive plant that has leaves with green nonscaly upper surfaces in summer and clusters of reddish, rounded berries in fall and early winter.

History and use. Native to Europe and Western Asia, a relatively recent (early 1900s) arrival in the upper part of the Southeast. Initially planted as a yard ornamental, for windbreaks, surface mine reclamation, and wildlife habitat. Widely escaped in the Western and Northeastern United States.

Distribution. Found infrequently as escaped plants from ornamental and surface mine plantings in forests and urban areas in VA and NC.

Summer

Chris Evans

Silktree, Mimosa

August

November

June

June

November

January

February

Albizia julibrissin Durazz. **ALJU** **Fabaceae**
Other common names: silky acacia, Japanese mimosa

Silktree, Mimosa **0345**

Plant. Deciduous, leguminous tree 10 to 50 feet (3 to 15 m) in height with single or multiple boles, smooth light-brown bark, feathery leaves, and showy pink blossoms that continually yield dangling flat pods during summer. Some pods persistent during winter.

Stem. Twigs slender to stout, lime green turning shiny grayish brown with light dots (lenticels). No terminal bud. Bark glossy, thin, light brown turning gray with raised corky dots and dashes.

Leaves. Alternate, bipinnately compound 6 to 20 inches (15 to 50 cm) long with 8 to 24 pairs of branches and 20 to 60 leaflets per branch, feathery and fernlike. Leaflets asymmetric, 0.4 to 0.6 inch (1 to 1.5 cm) long, dark green, with midvein nearer and running parallel to one margin. Margins entire.

Flowers. May to July (and sporadically to November). Terminal clusters at the base of current year twigs, each with 15 to 25 sessile flowers 1.4 to 2 inches (3.5 to 5 cm) long. Pompom-like with numerous filaments, bright pink feathery tufts with white bases. Fragrant.

Fruits and seeds. June to February. Legume pods in clusters, flat with bulging seeds, each pod 3 to 7 inches (8 to 18 cm) long, splitting in winter along the edges to release 5 to 10 oval seeds or disperse whole to float on water. Initially light green turning dark brown in fall and whitish tan in winter.

Ecology. Occurs on dry-to-wet sites and spreads along streambanks, preferring open conditions but also persisting in shade. Can form dense stands on abandoned farmland and home sites. Negatively impacts wildlife dependent on native vegetation. Seldom found above 3,000 feet (900 m). Forms colonies from root sprouts and spreads by abundant animal- and water-dispersed seeds. Seeds remain viable for many years. Nitrogen fixer.

Resembles honeylocust (*Gleditsia triacanthos* L.) and locusts (*Robinia* spp.), which have longer leaflets—1 to 2 inches (2.5 cm to 5 cm) long. **Also resembles** seedlings of partridge pea [*Chamaecrista fasciculata* (Michx.) Greene], an annual plant with once pinnately compound leaves, and littleleaf sensitive-briar (*Mimosa microphylla* Dryand.), a reclining legume with fine prickles.

History and use. A traditional ornamental introduced from Asia in 1745. Potential use for forage and biofuel.

Distribution. Found as scattered plants or dense infestations in forests and along highways and roadsides throughout the region.

June

Tallowtree, Popcorntree

June

November

September

Ted Bodner

September

September

September

June

Triadica sebifera (L.) Small **TRSE6** **Euphorbiaceae**
Synonym: *Sapium sebiferum* (L.) Roxb.
Other common name: Chinese tallowtree

Plant. Deciduous tree to 60 feet (18 m) in height and 3 feet (90 cm) in diameter, with broadly ovate leaves having extended tips. Dangling yellowish flower spikes in spring yield small clusters of 3-lobed fruit that split in fall and winter to reveal popcorn-like seeds.

Stem. Terminal clusters of flowers and fruits that drop in winter to result in whorled branching from lateral buds. Twigs lime green turning gray with scattered brownish dots (lenticels) later becoming striations. Numerous semicircular leaf scars becoming raised with age. Bark light gray and fissured. Sap milky.

Leaves. Alternate distinctively wide ovate with a rounded wide angled base and a short or long attenuated tip. Blades 2 to 3 inches (5 to 8 cm) long and 1.5 to 2.5 inches (4 to 6 cm) wide. Dark green with light-green mid- and lateral veins, turning yellow to red in fall. Hairless, lime-green petioles 1 to 3 inches (2.5 to 8 cm) long with 2 tiny glands on upper side of juncture between blade and petiole (requires magnification).

Flowers. April to June. Slender, drooping spikes to 8 inches (20 cm) long of tiny flowers. Yellowish-green sepals but no petals. Female flowers at base and males along the spike.

Fruit and seeds. August to January. Small terminal clusters of 3-lobed capsules (occasionally 4- to 5-lobed), each 0.5 to 0.75 inch (1.2 to 2 cm) across. Dark green in summer becoming black and splitting to reveal 3 white-wax coated seeds 0.3 inch (0.8 cm) long and 0.2 inch (0.5 cm) wide. Resemble popcorn and remain attached until winter.

Ecology. Invades streambanks, riverbanks, lakesides, and wet areas like ditches, as well as grassland prairies and upland sites. Thrives in both freshwater and saline soils. Shade tolerant, periodic-flood tolerant, and allelopathic. Increasing widely through ornamental plantings. Plants can produce viable seed by 3 years and remain reproductive for 100 years. Mature trees can produce 100,000 seeds per year. Spreading by bird- and water-dispersed seeds and colonizing by prolific surface root sprouts. Seeds remain viable in the leaf litter and soil for 2 to 7 years.

Resembles cottonwood (*Populus* spp.), which have leaves with toothed margins and flaking bark with fissured ridges.

History and use. Introduced from China to South Carolina and Georgia in the 1770s and then in significant numbers to the Gulf Coast in the early 1900s. Plantings for seed oil recommended by the U.S. Department of Agriculture during 1920 to 1940. Ornamentals still sold and planted. Waxy seeds traditionally used to make candles. Honey plant for beekeeping.

Distribution. Found in dense infestations in southeast TX; south and central LA; central and north FL; and the southern portions of AR, MS, AL, GA, and SC. Scattered infestations further north from ornamental plantings in cities and towns. The projected potential range includes all States within the region.

Tallowtree, Popcorntree **0994**

Thomas Ellis, Jr.

Tree-of-Heaven

August

July

February

July

August

August

October

Ailanthus altissima (P. Mill.) Swingle **AIAL** **Simaroubaceae**
Other common names: ailanthus, Chinese sumac, stinking sumac, paradise-tree, copal-tree

Tree-of-Heaven **0341**

Plant. Deciduous tree to 80 feet (25 m) in height and 6 feet (1.8 m) in diameter from a shallow root system, with long, pinnately compound leaves and circular glands under small lobes on leaflet bases. Strong unpleasant odor emitted from flowers and other parts when crushed, sometimes likened to peanuts or cashews.

Stem. Twigs stout, chestnut brown to reddish tan, smooth to velvety with light dots (lenticels) and large, heart-shaped leaf scars. Buds finely hairy, dome-shaped, and partially hidden by the leaf base. Branches light gray to dark gray, smooth and glossy, with raised dots becoming fissures with age. Bark light gray and rough with areas of light-tan fissures.

Leaves. Alternate, odd- or even-pinnately compound, 10 to 41 leaflets on 1 to 3 foot (30 to 90 cm) long, light-green to reddish-green stalks with swollen bases. Leaflets lanceolate and asymmetric and not always directly opposite, each 2 to 7 inches (5 to 18 cm) long and 1 to 2 inches (2.5 to 5 cm) wide. Long tapering tips and lobed bases with 1 or more glands beneath each lobe (round dots). Margins entire. Dark green with light-green veins above and whitish green beneath. Petioles 0.2 to 0.5 inch (5 to 12 mm) long.

Flowers. April to June. Large terminal clusters to 20 inches (50 cm) long of small, yellowish-green flowers, with 5 petals and 5 sepals. Male and female flowers on separate trees.

Fruit and seeds. July to February. Persistent clusters of wing-shaped fruit with twisted tips on female trees, 1 inch (2.5 cm) long. Single seed. Green turning to tan, then brown. Persist on tree for most of the winter.

Ecology. Rapid growing, forming thickets and dense stands. Both shade and flood intolerant and allelopathic. Colonizes by root sprouts and spreads by prolific wind- and water-dispersed seeds. Viable seed can be produced by 2- and 3-year-old plants. A mature female tree can produce up to 300,000 wind-dispersed seeds per year that can be distributed up to 330 feet (100 m) away.

Resembles hickories (*Carya* spp.), butternut (*Juglans cinerea* L.), black walnut (*J. nigra* L.), and sumacs (*Rhus* spp.), which have pinnately compound leaves but no glands at leaflet bases. Hickories distinguished by braided bark, butternut and black walnut by their ridged mature barks, and all have large nuts. Sumacs often in a shrub shape, red or winged leaf stalks, and terminal conical flower and seed clusters.

History and use. Introduced in 1784 from Europe, although originally from Eastern China. Ornamental widely planted in cities due to pollution and drought tolerance.

Distribution. Found throughout the region with dense infestations in central VA, KY, TN, NC, and SC. Scattered occurrences southward from ornamental plantings.

July

Trifoliate Orange, Hardy Orange

October

July

March

May

March

March

March

Poncirus trifoliata (L.) Raf **POTR4** **Rutaceae**

Trifoliate Orange, Hardy Orange

Plant. Deciduous, small tree or shrub to 20 feet (6 m) in height with tufts of trifoliate (3-leaflet) leaves on densely packed stems with dangerously sharp axillary thorns. White flowers in spring cover plants to yield abundant small, fuzzy green orange-like fruit that turn yellow in fall. Root sprouts abundant around stems.

Stem. Twigs flattened, glossy green and hairless, turning yellowish in drought and winter, becoming stout, chestnut brown to reddish tan, smooth to velvety with light dots (lenticels), heart-shaped leaf scars and sharp thorns jutting outward to 2 inches (5 cm). Buds finely hairy, dome-shaped, and partially hidden by the leaf bases. Branches dark gray with lengthwise lighter stripes and intervening green stripes that become an intricate braided network on the bark. Basal sprouts vinelike and green, climbing up through the crown into surrounding trees.

Leaves. Alternate or tufts of trifoliate (3-leaflet) leaves in the axils of thorns, appearing at or just after the time of flowering in the spring. Leaflets unequal in size with the terminal 1 to 2.5 inches (2.5 to 6.4 cm) long and 0.5 to 1 inch (1.2 to 2.5 cm) wide, obovate to elliptic, while the lateral leaflets are similar but smaller. Dark green and hairless, becoming yellowish in the fall. Blades merge into the winged leaf stalk (sometimes not winged).

Flowers. March to early May. Showy clusters of white, 5-petaled flowers, 1.5 to 2 inches (3.8 to 5 cm) across that cover trees early on previous year's branches. In the center of the flower are 8 to 10 projecting stamens that enclose a yellow, hairy stigma.

Fruit and seeds. July to October and rarely persisting into winter. A hairy, bitter orange, green turning yellow to golden when ripe, up to 1.5 inches (3.8 cm) wide. Pulp is minimal and the multiple seeds are viable.

Ecology. Moderate initial growth rate becoming rapid with establishment, forming dense, impenetrable thickets and stands. Prefers open areas or edges and acid, well-drained soils. Colonizes by basal sprouts and spreads by prolific animal-dispersed seeds.

Resembles osage-orange [*Maclura pomifera* (Raf.) C.K. Schneid.], which has similar thorny stems, but larger, spherical yellowish fruit that are not hairy and leaves that are not trifoliate.

History and use. Introduced as an ornamental in the 1850s from China and Korea. Fruits are useful for extraction of acidic juice high in vitamin C. Used as a hearty rootstock for grafted citrus. Historically planted as a thorny hedge to confine livestock.

Distribution. Found in scattered dense infestations in TX, OK, AR, LA, and GA with occasional occurrences elsewhere throughout the region except KY.

Tungoil Tree

Vernicia fordii (Hemsl.) Airy-Shaw **VEFO** **Euphorbiaceae**
Synonym: *Aleurites fordii* Hemsl.
Other common name: Chinese wood-oil tree

Tungoil Tree **0995**

Plant. Deciduous tree (leaves fall with frost) to 40 feet (12 m) in height having a rounded crown with many alternate branches and basal sprouts. Sap is milky white. Leaves heart shaped, some with rounded sinuses, and long petioles with a pair of conspicuous dark glands where joining the blade. Clusters of showy, white and rose, flaring flowers in spring yield large spherical "nuts" in the fall. **Caution: Leaves and nuts are poisonous.**

Stem. Twigs moderately stout, often radiating outward in numbers from a swollen branch knot that terminates each year's growth. The branch knot has multiple leaf and fruit stem scars along with numerous protruding light dots. Twigs light green, becoming mottled with a silvery-gray film and turning grayish to gray tan with whitish dots (lenticels) that turn to faint lengthwise stripes that increase with age. Buds overlapping, maroon, with multiple, leafy scales. Pith white and spongy. Leaf scars circular when whorled or oval when stacked together. Larger scars reveal 8 vascular bundle scars. Bark light tan to light gray, tight, covered with corky dots.

Leaves. Alternate but more clustered at twig tips, heart shaped, 3 to 14 inches (7.5 to 35 cm) long with 1 tip, or lobed with deep sinuses and 3 to 5 pointed tips and a cordate base. Two rounded, dark reddish-maroon glands occur where the petiole joins the blade. Glossy and dark green above with 5 prominent light-green veins radiating from the base, whitish silvery beneath. Petioles 3 to 6 inches (8 to 15 cm) long, green with a maroon tinge. Leaves turning yellow in the fall with showy maroon petioles and lower veins.

Flowers. March to April. Large, terminal branched clusters of separate male and female flowers appearing before leaves to cover the tree. Widely flared flowers about 1 inch (2.5 cm) wide having 5 to 7 brilliant white petal lobes splashed with red to maroon within the throat radiating outward in lines, and protruding yellow floral parts. Stalks to 6 inches (15 cm) long, smooth, red to orange, the same color as the sepals.

Fruit and seeds. September to November. Large, spherical, woody nuts (drupes), 2 to 3 inches (5 to 8 cm) wide, dark green turning maroon to finally brown, dropping whole to the ground or water to split into 3 to 7 wedged-shaped, fibrous sections each with a large brownish nut, about 1 inch (2.5 cm) long.

Ecology. Rapid growing in moist and well-drained soils, forming dense stands. When first introduced, fruit could not withstand freezing temperatures, until crop breeding yielded frost-hardy varieties. Colonizes by stump sprouts and spreads by animal- and water-dispersed seeds. Viable seed can be produced at 3 years.

Resembles paper mulberry [*Broussonetia papyrifera* (L.) L'Hér. ex Vent], southern catalpa (*Catalpa bignonioides* Walt.), northern catalpa [*C. speciosa* (Warder) Warder ex Engelm.], and princesstree [*Paulownia tomentosa* (Thunb.) Sieb. & Zucc. ex Steud.], which have similar shaped leaves but are velvety or rough hairy and have no petiole glands. **Also resembles** Chinese parasoltree [*Firmiana simplex* (L.) W. Wight], which has similar shaped leaves with sinuses but has no glands and no milky sap.

History and use. Initially introduced in 1905 from China with further introductions to the lower Gulf Coast States for use in the tungoil industry, which collapsed by the 1950s due to freezes, hurricanes, and offshore competition. Sparingly planted as an ornamental for the showy flowers.

Distribution. Found in dense infestations in LA and the southern portions of MS, AL, and GA, and north FL. Scattered occurrence northward from ornamental plantings.

August October August

Autumn Olive

PA Dept. Conserv. Nat. Res. - Forestry Archive

Elaeagnus umbellata Thunb. **ELUM** **Elaeagnaceae**
Other common names: Russian olive, oleaster

Autumn Olive **2038**

Plant. Tardily deciduous, bushy and leafy shrub, 3 to 20 feet (1 to 6 m) in height, with scattered thorny branches. Leaves silver scaly beneath, with many red berries in fall.

Stem. Twigs slender and silver scaly, spur twigs common, with some lateral twigs becoming pointed, like thorns. Branches and main stems glossy and gray green with scattered thorns and many whitish dots (lenticels), becoming light gray to gray brown with age and eventually fissuring to expose light-brown inner bark.

Leaves. Alternate, elliptic, 2 to 3 inches (5 to 8 cm) long and 0.8 to 1.2 inches (2 to 3 cm) wide. Margins entire and wavy. Bright green to gray green above with silver scaly midvein and densely silver scaly beneath. Petioles short and silvery.

Flowers. February to June. Axillary clusters, each with 5 to 10 tubular flowers with 4 lobes. Silvery white to yellow. Fragrant.

Fruit and seeds. August to November. Round, juicy drupe 0.3 to 0.4 inch (8 to 10 mm) wide containing 1 nutlet. Red and finely doted with silvery to silvery-brown scales.

Ecology. Prefers drier sites. Shade tolerant. Spreads by animal-dispersed seeds and found as scattered plants in forest openings and open forests, eventually forming dense stands. A nonleguminous nitrogen fixer.

Resembles silverthorn or thorny olive (*E. pungens* Thunb.) and Russian olive (*E. angustifolia* L.). Silverthorn is an evergreen that has brown scaly and hairy twigs, flowers in late fall, and oval reddish-silver, scaly drupes in spring. Russian olive rarely occurs and has silver scaly twigs and leaves, leaves longer and more linear, flowers in early summer, and many yellow olives in fall and winter. **Also resembles** plum (*Prunus* spp.) when fruit is present, although plum leaves are not silvery beneath and fruit is much larger.

History and use. Introduced from China and Japan in 1830. Planted for wildlife food and surface-mine reclamation.

Distribution. Found throughout the region with dense infestations more frequent in VA, KY, SC, and GA.

June

Bush Honeysuckles

Sweet Breath of Spring
September

Amur

Amur
Spring
Warner Park

Sweet Breath of Spring
September

Amur
December

Amur
Spring
John Schwegman

Tatarian
May
Patrick Breen

Amur
December

Amur
December

Tatarian Honeysuckle, *L. tatarica* L. LOTA **Caprifoliaceae**
Amur Honeysuckle, *Lonicera maackii* (Rupr.) Herder **LOMA6**
Morrow's Honeysuckle, *L. morrowii* A. Gray **LOMO2**
Sweet-breath-of-spring, *L. fragrantissima* Lindl. & Paxton **LOFR**
Other common names: winter and fragrant honeysuckle, January jasmine
Bell's Honeysuckle, *L. x bella* Zabel **LOBE**
(hybrid Morrow's and Tatarian)

Plant. Tardily deciduous, upright, arching-branched shrubs to small trees. Amur to 30 feet (9 m) in height and spindly in forests, Morrow's to 6.5 feet (2 m) in height, Tatarian and sweet-breath-of-spring to 10 feet (3 m) in height, and Bell's to 20 feet (6 m) in height. Much branched and arching in openings, multiple stemmed, dark-green opposite leaves, showy white to pink or yellow flowers, and abundant orange to red berries.

Stem. Opposite branched, light tan with braided-strand appearance. Bark often flaking. Older branches hollow.

Leaves. Opposite in 2 rows, ovate to oblong with rounded bases, 1.2 to 4 inches (3 to 10 cm) long. Persistent into winter. Margins entire. Amur tapering to a long slender tip; Bell's to a medium tapering tip; and others with short pointed tips. Morrow's with wrinkled upper surface and both Amur and Bell's soft-hairy lower surface, others with hairless leaves. Petioles 0.1 to 0.4 inch (2.5 to 10 mm) long.

Flowers. February to June. Axillary, bracted short-stemmed clusters, each with 1 to several white to yellow (some pink to red) flowers. Petals tubular flaring to 5 lobes in 2 lips (upper lip 4-lobed, lower lip single-lobed). Five extended stamens. Fragrant.

Fruit and seeds. June to March. Abundant spherical, glossy berries paired in leaf axils, each 0.2 to 0.5 inch (6 to 12 mm). Green becoming pink and ripening to red (sometimes yellow or orange). Usually persistent into winter and sometimes spring.

Ecology. Often forms dense thickets in open forests, forest edges, abandoned fields, pastures, roadsides, and other open upland habitats. Relatively shade tolerant. Colonize by root sprouts and spread by abundant bird- and other animal-dispersed seeds. Seeds long lived in the soil.

Resemble the woody vine, Japanese honeysuckle (*L. japonica* Thunb.) as far as leaves and flowers. **Also resemble** the native shrub American fly honeysuckle (*L. canadensis* Bartr. ex Marsh.), which has hairy-margined leaves, blue fruit, and is found at high elevation in mountains. **Also resemble** the native bush honeysuckles (*Diervilla* spp.), which have similar leaves but terminal flowers in cymes and capsules for fruit.

History and use. All introduced from Asia in the 1700s and 1800s, and used as ornamentals and wildlife plants. Some still sold.

Distribution. Found throughout the region with dense infestations in KY, TN, central VA, and north AR.

Bush Honeysuckles **2105**

Invasive SHRUBS

Amur honeysuckle November

Warner Park

Chinese / European / Border / California Privet

Ted Bodner

Ted Bodner

Chinese privet shown in all images

Ted Bodner

Chinese Privet, *Ligustrum sinense* Lour. LISI **Oleaceae**
European Privet, *L. vulgare* L. LIVU
Border Privet, *L. obtusifolium* Sieb. & Zucc. LIOB
California Privet, *L. ovalifolium* Hassk. LIOV

Plant. Thin, opposite-leaved, evergreen, thicket-forming shrubs to 30 feet (9 m) in height that are multiple stemmed and leaning to arching with long, leafy branches. Much used as border shrubs. Chinese privet is one of the most widely invasive plants in the South, while other three are less frequent. It is difficult to distinguish between Chinese and European, with probable blurring due to hybridization. **Caution: Fruit (berries) is poisonous.**

Stem. Opposite, long slender branching that increases upward with shorter twigs projecting outward at near right angles. Brownish gray turning gray green and hairy or not with light dots (lenticels). Leaf scars semicircular with 1 bundle scar. Bark light gray to brownish gray and slightly rough (not fissured).

Leaves. Thin and opposite in 2 rows at near right angle to stem. Chinese and European: ovate to elliptic with rounded tip (often minutely indented), 0.8 to 1.6 inches (2 to 4 cm) long and 0.4 to 1.2 inches (1 to 3 cm) wide, hairless beneath. Lustrous green above and pale green beneath with Chinese having a hairy midvein beneath. Border: elliptic-oblong, 1 to 2.2 inches (2.5 to 6 cm) long and 0.3 to 1 inch (0.8 to 2.5 cm) wide, hairless green above and hairy beneath. California: oval to elliptic with wedge-shaped base, 1.2 to 2.4 inches (3 to 6 cm) long and half as wide, lustrous green above and yellow green beneath and hairless. Margins entire. Petioles 0.04 to 0.2 inch (1 to 5 mm) long. Leaves usually persistent during winter, while California privet is deciduous northward.

Flowers. April to June. Abundant, terminal and upper axillary clusters on short branches forming panicles of white to cream flowers. Corolla 4-lobed, to 0.6 inch (1.8 cm) long, with stamens extending or within the corolla. Fragrance causing sinus irritation in many people, with California being the most unpleasant.

Fruit and seeds. July to March. Dense ovoid drupes hanging or projecting outward, 0.2 to 0.3 inch (6 to 8 mm) long and 0.16 inch (4 mm) wide, containing 1 to 4 seeds. Pale green in summer ripening to dark purple and appearing almost black in winter.

Ecology. Aggressive and troublesome invasives, often forming dense thickets, particularly in bottomland forests and along fencerows, thus gaining access to forests, fields, and right-of-ways. Shade tolerant. Colonize by root sprouts and spread widely by abundant bird- and animal-dispersed seeds.

Resemble Japanese privet (*L. japonicum* Thunb.) and glossy privet (*L. lucidum*), which have larger leaves and are further described in this book. **Also resemble** native swampprivet (*Forestiera* spp.), which have leaves on short twigs sparse axillary flowers, and few fruit.

History and use. Introduced from China and Europe in the early to mid-1800s. Traditional southern ornamentals. Variegated cultivars of Chinese privet are widely planted in the coastal South. Deer browse Chinese privet sprouts.

Distribution. Chinese and European privets are found throughout the region in dense infestations along highways and roadside margins, parks and preserves, bottomland and interior forests with most frequent infestations in MS, AL, GA, and SC. Border privet in KY, VA, NC, and TN. California privet in KY, VA, NC, AL, and FL, at present.

Chinese / European / Border / California Privet 2103

May

Hen's Eyes, Coral Ardisia

David J. Morehead

Ron Lance

Ardisia crenata Sims **ARCR80** **Myrsinaceae**

Other common names: spiceberry, coralberry, Christmas berry

Plant. Evergreen, erect shrub, 2 to 6 feet (0.6 to 1.8 m) in height with short stems or multi-stemmed bushy clumps. Shiny green leaves with distinct thickened, wavy margins, drooping white to pink axillary flowers, and dangling, bright red berries in fall and through winter. No rhizomes.

Stem. Twigs light green and shiny, projecting alternately outward from light brown, erect stems, becoming increasingly rough, with grayish bark. Leaf and stem scars broadly V-shaped with a raised bud at top.

Leaves. Alternate, 4 to 8 inches (10 to 20 cm) long, elliptic to narrowly lanceolate or oblanceolate with a pointed tip, leathery with scalloped crenate margins (thus the scientific name, *crenata*) and raised callused notches. Shiny and dark green above with a paler midvein and pale green beneath, tapering to a short winged petiole.

Flowers. April to October. Axillary clusters dangling below the leaves, with green to red stems, conical buds mixed with flowers having 4 to 5 white to pink petals, yellow centers and eventually an extending style.

Fruit and seeds. November to March. Abundant, spherical 1-seeded drupes, 0.2 to 0.3 inch (6 to 8 mm) wide, hanging down in fanned clusters often jutting outward on lower branch ends, green then ripening through shades of coral to finally bright scarlet.

Ecology. Forms infestations in partial shade or full shade and grows best in moist, well-drained soils. Forms dense infestations to shade out ground flora. Spreads by animal-dispersed seed and produces fruit within 2 years.

Resembles only the related shoebutton (*A. elliptica* Thunb.), which does not have wavy margins and has fruit ripening to black, and only invasive in FL wetlands at present.

History and use. Native to Japan and Northern India and introduced into FL in 1900 as an ornamental. Still being sold and planted in the Southeast, and worldwide by internet sales.

Distribution. Found as dense infestations throughout LA and FL and recently found in south GA.

Hen's Eyes, Coral Ardisia **FL11**

February

Japanese/Glossy Privet

Glossy privet

June

Glossy privet

October

Japanese privet

July

Ted Bodner

Glossy privet

June

Glossy privet

December

Glossy privet

June

Japanese Privet, *Ligustrum japonicum* Thunb. **LIJA** **Oleaceae**
Glossy Privet, *L. lucidum* Ait. f. **LILU2**

Plant. Opposite, thick-leaved evergreens to 20 feet (6 m) in height for Japanese privet and 35 feet (10 m) in height for glossy privet, with spreading crowns, conical clusters of white flowers in spring, and green to purple-black fruit in summer to following spring.

Stem. Twigs hairless and pale green becoming brownish to reddish tinged. Branches opposite and brownish gray with many raised corky dots (lenticels). Bark light gray and smooth except for scattered horizontal, discontinuous ridges.

Leaves. Opposite, leathery, ovate to oblong, bases rounded and tips blunt or tapering, often with a tiny sharp tip. Two to 4 inches (5 to 10 cm) long and 1 to 2 inches (2.5 to 5 cm) wide. Margins entire and often yellowish rimmed, turned upward with glossy privet and slightly rolled under with Japanese privet. Upper blades lustrous dark green with 6 to 8 pairs of light-green veins with glossy privet and 4 to 6 pairs of indistinct veins that protrude slightly from light green lower surfaces with Japanese privet. Petioles 0.4 to 0.8 inch (1 to 2 cm) long and often reddish tinged for glossy privet and 0.2 to 0.4 inch (6 to 12 mm) long and light green for Japanese privet.

Flowers. April to June. Loosely branching, terminal- and upper-axillary, conical clusters of many small white 4-petaled flowers. Fragrant.

Fruit and seeds. July to February. Conical-shaped, branched terminal clusters of ovoid drupes, each 0.2 to 0.5 inch (5 to 12 mm) long and 0.2 inch (5 mm) wide. Pale green in summer ripening to blue black in winter.

Ecology. Single plants or thicket-forming, occurring in the same habitats as Chinese privet, but generally not as abundant, depending upon location. Invade both lowland and upland habitats, but usually more prevalent in lowlands. Shade tolerant. Colonize by root sprouts and spread by abundant bird- and animal-dispersed seeds.

Resembles Chinese privet (*L. sinense* Lour.), which has smaller and thinner leaves and is further described in this book. **Also resembles** red tip or photinia (*Photinia* spp.). and Carolina laurel cherry (*Prunus caroliniana* Ait.), which have similar evergreen, but alternate, leaves with finely toothed margins.

History and use. Introduced from Japan and Korea in 1845 and 1794, respectively. Widely planted as ornamentals and escaped.

Distribution. Found in dense infestations and as scattered escaped plants throughout the region.

Japanese/Glossy Privet **2104**

Glossy privet June

Japanese Barberry

April

April

November

October

Barry Rice

April

April

May

Leslie J. Mehrhoff

Berberis thunbergii DC. BETH **Berberidaceae**
Synonym: _B. thunbergii_ DC. var. _atropurpurea_ Chenault

Japanese Barberry

Plant. Tardily deciduous, compact and spreading shrub, 2 to 4 feet (60 to 120 cm) in height and slightly wider, occasionally 6 to 8 feet (1.8 to 3 m) in height. Multiple alternate, slender and brown stems bear single slender spines at nodes, along with tufts of small paddle-shaped leaves ranging from green to reddish and turning orange and golden to crimson in early winter. Tiny, dangling clusters of white flowers in spring yield red, fleshy, oblong berries. Roots shallow, yellow inside as are stems. The plant traits are variable due to numerous cultivars that are still being sold and planted.

Stem. Twigs slender, reddish brown to gray, ridged to varying degrees, a single sharp spine to 0.6 inch (1.5 cm) long at each node, below a stubby, conelike stacked bud (evident in winter). Stems sometimes wavy in appearance owing to numerous alternate nodes. Alternate branching with bark becoming irregularly furrowed, light gray to tan, with scattered, stacked buds remaining.

Leaves. Alternate in tight clusters, spatulate (paddle shaped) with entire margins and bases narrowing to the stem, with rounded or faintly pointed tips, variable due to cultivars, 0.5 to 1.5 inches (1.2 to 3.7 cm) long. Blue green to green to reddish above and pale to whitish below, turning orange and golden to red and crimson in winter. Appear early and remain late.

Flowers. April to May. Dangling singly or in small clusters from most nodes, 6 white to yellowish-white to yellow petals, 0.2 to 0.3 inch (5 to 6 mm) wide with yellow centers.

Fruit and seeds. May and maturing by October, remaining until March. Dangling, egg-shaped berries, red and shiny, fleshy, 0.3 to 0.4 inch (8 to 11 mm) long. Each containing 1 brown, pitted seed, oblong, 0.08 to 0.20 inch (2 to 5 mm) long.

Ecology. Forms dense infestations under forest canopies, but prefers partial shade of edges, to exclude other plants. Seeds are dispersed by many wildlife species. Deer do not browse this species, resulting in release from surrounding browsed vegetation. Infestations intensify by root sprouts and rooting of drooping stems. Seeds remain viable for up to 10 years in the soil.

Resembles both the nonnative invasive common or European barberry (_B. vulgaris_ L.) and the native American barberry (_B. canadensis_ Mill.), while both have finely bristled leaf margins. **Also resembles** a rare escaped wintergreen barberry (_B. julianae_ C.K. Schneid.) in the mountains that has leathery, evergreen leaves.

History and use. Native to Japan and introduced into the U.S. as an ornamental in the mid-1870s. Numerous cultivars widely sold and planted with varying amounts of fruit production.

Distribution. Found as scattered plants and a range of infestation densities in GA, SC, NC, TN, KY, and VA.

Leslie J. Mehrhoff

Steven T. Manning

Steven T. Manning

Japanese Knotweed

July

Steven T. Manning

September

June

July

September

Leslie J. Mehrhoff

September

October

June

November

Steven T. Manning

Polygonum cuspidatum Siebold & Zucc **POCU6** **Polygonaceae**
Synonyms: *Fallopia japonica* (Houtt.) Dcne., *Reynoutria japonica* Houtt.
Other common names: fleeceflower, Mexican bamboo

Japanese Knotweed

Plant. Tall perennial, herbaceous shrub 3 to 12 feet (1 to 3.5 m) high, freely branching in dense, often clonal, infestations. Reddish stems, hollow and jointed like bamboo, survive only 1 season while rhizomes up to 65 feet (20 m) long survive decades. Alternate leaves appear in spring on new sprouts, ovate with pointed tips and flat bases. In late summer, sprays of tiny, white flowers emerge along stalks at leaf axils, yielding abundant tiny-winged seeds. Dead plants remain upright or leaning during winter and burn hot to pose a severe fire hazard.

Stem. Round, reddish brown to mottled with green, to about 1 inch (2.5 cm) in diameter, resembling bamboo although not woody, smooth with scattered to many tiny dots (scales), often ridged, having hollow internodes and swollen solid nodes with membranous sheaths clinging to the base of the nodes. Profuse red to green, slender branches grow upward and outward, and some drooping to form dense entanglements.

Leaves. Alternate and broadly ovate to oblong ovate, 4 to 6 inches (10 to 15 cm) long and 3 to 4 inches (8 to 10 cm) wide with distinctly pointed tips and straight wedge bases. Smooth and bright green above with whitish indented veins and dull green beneath with protruding veins. Petioles reddish, 0.5 to 1 inch (1.2 to 2.5 cm) long. Leaves turn bright yellow in fall.

Flowers. May to September. Terminal and axillary, branched sprays (racemes) 3 to 6 inches (8 to 15 cm) long, covered with tiny 5-petaled (sepaled) white to greenish flowers all having 3 styles and 8 to 10 stamens. Functionally male or female flowers can occur on different plants or within a raceme.

Fruit and seeds. August to November. Many dangling, winged fruit that can contain 1 triangular, shiny nutlet. Viability apparently variable but can be quite high in some stands.

Ecology. Tolerates a wide range of growing conditions from full sun to shade, to high salinity and drought, while it prefers wet soils in low places or along streams and rivers. Spreads along streams by stem and rhizome fragments and seeds to dominate extensive riparian habitat. Also spreads along highways and roads by similar means through maintenance mowing. A serious threat to native habitats since the dense infestations exclude all other plants and animals.

Resembles the nonnative invasive giant knotweed (*P. sachalinense* F. Schmidt ex Maxim.) a larger plant with greenish flowers and cordate leaves with tapering points, currently found in KY, VA, TN, NC, and LA. These invasives hybridize in the Northeast (NE).

History and use. Introduced from China, Japan, and Taiwan in the late 1800s as an ornamental. An invasive in many parts of the world.

Distribution. Found in VA, KY, TN, and NC with scattered occurrences elsewhere except in OK, TX, and FL.

July

Japanese Meadowsweet

September

Great Smoky Mtns. NP Res. Mgmt. Archive

June

September

September

September

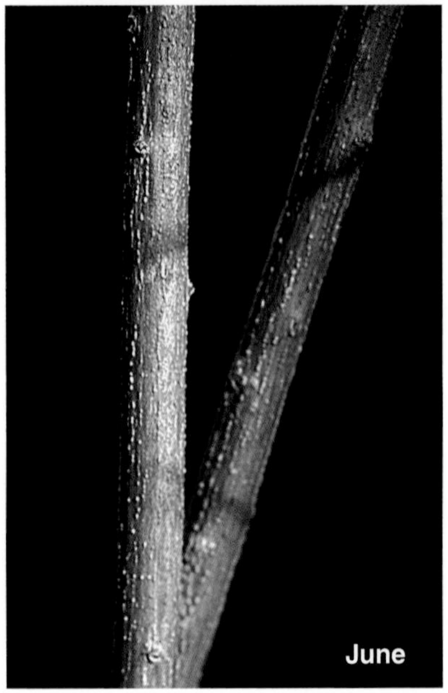

June

September

Spiraea japonica L. f. **SPJA** **Rosaceae**
Other common names: Japanese spiraea, maybush

Japanese Meadowsweet

Plant. Deciduous, erect shrub to 6 feet (1.8 m) high with multiple stems and alternate branches, slender and brown, intertwining within older shaded plants or arching outward on hillside infestations. Small, alternate, lanceolate leaves with irregular serrate margins and flat-topped clusters of tiny, rose-pink flower heads that festoon branch tips and turn into crowded clusters of lustrous brown seed capsules in midsummer. Plant traits are variable due to numerous escaped cultivars.

Stem. Twigs initially whitish green, slender and wiry, usually quite hairy, jutting upward, often terminated in flower or fruit clusters. Stems becoming brown to reddish brown, round in cross section, with increasing grayish to whitish lengthwise stripes of dotted cork and protruding circular leaf and branch scars surrounded by corky deposits. Multiple basal sprouts on older plants intertwine for support.

Leaves. Alternate, thin, elliptic to lanceolate, 1 to 3 inches (2.5 to 7.6 cm) long and less than half as wide, tips are pointed and bases wedge shaped, margins irregularly and sharply serrated but less so towards the base. Blue green to green above and lighter beneath. A short petiole connects to a pronounced whitish-green midvein with about 10 lateral, curved veins that end at serrate tips and are hairy beneath.

Flowers. June to September. Terminal on new growth, multibranched, flat-topped clusters (corymbs) of many red to rosy, domed flower buds that open into tiny, whitish-pink to rose (rarely white) flowers, 0.2 inch (5 mm) wide with 5 petals and an extended mist of anthers. Sepals and stalks hairy.

Fruit and seeds. July to November. Flat-topped heads of tiny star-shaped clusters of 5 smooth and lustrous, tan to brown capsules that split on top at varying times to release 1 minute golden seed, 0.09 to 0.1 inch (2 to 2.4 mm) long.

Ecology. Forms dense infestations of entangled stems and branches and produces abundant foliage to exclude other plants and impact animal habitat. Seeds are dispersed by gravity, water, and soil movement during road maintenance. Populations occur along streams, roads, and adjacent disturbed sites and move into forest gaps and understories. Infestations intensify by abundant basal sprouting. Tolerant of a wide variety of growing conditions.

Resembles several native and nonnative spiraeas, but is unique in the flat-topped, pink to pink-rose flower clusters and brown fruit clusters, the hairy branchlets and flowers, and lanceolate leaves. It is the species most often found growing in dense infestations. The other nonnative spiraea species have yet to become so problematic.

History and use. Introduced from Japan, Korea, and China as an ornamental about 1870. Several cultivars still sold and planted by unsuspecting gardeners.

Distribution. Found as scattered dense infestations in VA, KY, TN, NC, SC, and north GA.

September

Leatherleaf Mahonia

April

Nancy Loewenstein

February

Nancy Loewenstein

April

September

February

Nancy Loewenstein

April

Nancy Loewenstein

Mahonia bealei (Fortune) Carrière **MABE2** **Berberidaceae**
Synonyms: *Berberis bealei* (Fortune) Carrière
Other common names: Beale's barberry, Beale's Oregon-grape

Plant. Evergreen shrub up to 10 feet (3 m) in height and branching to 4 to 8 feet (1.2 to 2.4 m) wide, erect and gangly or multi-stemmed from a pronounced root crown (with shallow roots). Leathery, odd-pinnately compound leaves radiate outward from the stem on long stalks with spiny, holly-like leaflets. Terminal, radiating stems of fragrant, yellow flowers in late winter to spring yield robin's egg blue fruit covered by whitish wax, maturing to bluish black. Wood and inner roots bright yellow.

Stem. Terminal stem growth comprised of crowded and overlapping broad leaf bases, light green or purple on seedlings, soon developing thin, tan to gray, fissured bark. The clasping leaf bases remain greenish and spaced at intervals along stout stems.

Leaves. Odd-pinnately compound, over 1 foot (30 cm) long on purplish stalks, stiff and spiraling out at intervals from the main stem with 9 to 13 leathery leaflets, 1 to 4 inches (2.5 to 10 cm) long. Leaflets with 5 to 7 extremely sharp marginal spines, holly-like, with no petioles and the terminal leaflet being largest. Lustrous green above with a lighter midvein and pale green beneath. Seedlings initially have simple heart-shaped leaflets on long petioles with many spines around the margins and often very white waxy beneath.

Flowers. January to April. Plants are topped with 6 to 12 unbranched, bluish to purplish stems with lateral dangling yellow fragrant flowers, opening from base to tip.

Fruit and seeds. March to August. Many fleshy-skinned, egg-shaped berries, 0.4 to 0.7 inch (1 to 1.8 cm) long, dangle from bracts on a purplish stalk. Berries white waxy coated, light green turning robin's egg blue and ripening purplish black, each with 2 to 3 seeds.

Ecology. Moderate in growth rate in a full range of soil textures and light conditions, but prefers moist soils. Tolerates cold, however, young plants can be damaged by late frosts. Pollinated by insects. Colonizes by basal sprouts and spreads by many bird-dispersed seeds from ornamental and escaped plants. Seed from ripe fruit can immediately germinate. Hybridizes with other mahonias.

Resembles native subshrubs hollyleaved barberry [*M.aquifolium* (Pursh) Nutt.] and Cascade barberry [*M. nervosa* (Pursh) Nutt.], which are small, low growing, and have 6 to 13 spiny teeth around their leaf margins.

History and use. Introduced in 1845 from China, Japan, and Taiwan as an ornamental. Still being sold and planted by unsuspecting gardeners.

Distribution. Found as scattered plants and new infestations in AL, GA, FL, SC, NC, and VA.

Leatherleaf Mahonia

Nancy Loewenstein

Nonnative Roses

Macartney rose — September

Macartney rose — May

Multiflora rose — May

Cherokee rose — April

Macartney rose — September

Multiflora rose — October

Cherokee rose — April

Multiflora rose — December

Multiflora rose — October

Multiflora rose — June

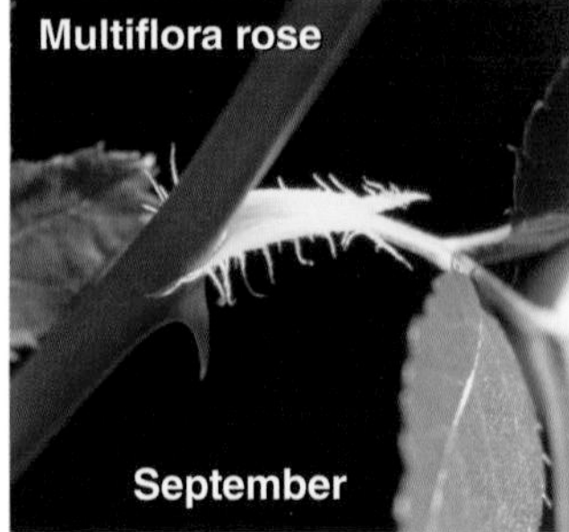
Multiflora rose — September

Multiflora rose, *Rosa multiflora* Thunb. ex Murr. ROMU **Rosaceae**
Macartney rose, *R. bracteata* J.C. Wendl. ROBR
Cherokee rose, *R. laevigata* Michx ROLA
Other nonnative roses, *Rosa* spp.

Plant. There are 21 nonnative rose species loose in southern ecosystems and only 8 natives. These descriptions are for some of the most invasive roses, but are not all inclusive. Evergreen (except multiflora) erect climbing, arching, or trailing shrubs to 10 feet (3 m) in height or length. Clump forming. Pinnately compound leaves, frequent recurved and straight prickles, clustered or single white flowers in early summer, and red rose hips in fall to winter.

Stem. Long arching or climbing by clinging using recurved or straight thorny prickles. Green with linear horizontal leaf and branch scars at nodes. Flower buds of multiflora often red in winter. Bark dark brown with streaks of light brown or green.

Leaves. Alternate, odd-pinnately compound with 3 to 9 elliptic to lanceolate leaflets, each 1 to 3 inches (2.5 to 8 cm) long. Margins finely and sharply serrate. Leafstalk bases clasping, channeled, and often bristled on margins, with toothed hairs.

Flowers. April to June. Terminal or axillary branched clusters or single flowers. Five white, pink, or red petals with many yellow anthers in center.

Fruit and seeds. July to December. Rose hip, spherical, and fleshy, 0.25 to 0.4 inch (0.6 to 1 cm). Green to yellow and ripening to glossy red.

Ecology. Form small-to-large infestations often climbing up into trees. Multiflora widely planted and often spreading along right-of-ways and invading new forests and forest margins. Colonize by prolific sprouting and stems that root, and spread by animal-dispersed seeds.

Resemble native Carolina rose (*R. carolina* L.), swamp rose (*R. palustris* Marsh.), and climbing rose (*R. setigera* Michx.), all of which have pink flowers in spring and nonbristled leafstalk bases, but none forming extensive infestations except swamp rose in wet habitat.

History and use. Introduced from Asia. Traditionally planted as ornamentals, livestock containment, and wildlife habitat. Multiflora widely planted for "living fences" or screening. Rose hips collected by herbalists for their high vitamin C content.

Distribution. Found throughout the region except in FL with dense and frequent infestations in TX, AR, MS, AL, TN, KY, VA, NC, and west SC.

Nonnative Roses **2160**

Multiflora rose

May

Sacred Bamboo, Nandina

December

August

September

May

May

September

Fred Nation

December

Nandina domestica Thunb. **NADO** **Berberidaceae**
Other common name: heavenly bamboo

Sacred Bamboo, Nandina

Plant. Evergreen erect shrub to 8 feet (2.5 m) in height, with multiple bushy stems that resemble bamboo, glossy bipinnately compound green (or reddish) leaves, white to pinkish flowers in terminal clusters, and bright-red berries in fall and winter.

Stem. Large compound leaves resembling leafy branches with overlapping sheaths clasping the main stem, woody leafstalk bases persisting as stubby branches. Stubby branches whorled alternately up the stem and tightly stacked near terminals for a given year's growth. The overlapping sheaths on the main stem give the appearance of bamboo (thus, the common name). Stem fleshy and greenish gray near terminal, becoming woody barked and tan to brown with fissures towards the base. Wood bright yellow.

Leaves. Alternate spiraled, at branch tips, bipinnately compound on 1.5 to 3 feet (0.5 to 1 m) long slender, often reddish tinged leafstalks, with joints distinctly segmented. Leafstalk bases clasping stems with a V-notch on the opposite side of attachment. Nine to 81 nearly sessile leaflets, lanceolate to diamond shaped, 0.5 to 4 inches (1.2 to 10 cm) long and 0.4 to 1.2 inches (1 to 3 cm) wide. Glossy light green to dark green sometimes red tinged or burgundy.

Flowers. May to July. Terminal (or axillary) panicles of several hundred flowers, 4 to 10 inches (10 to 25 cm) long. Pink in bud, opening to 3 (2 to 4) lanceolate deciduous petals, white to cream, with yellow anthers 0.2 to 0.3 inch (6 to 8 mm) long. Fragrant.

Fruit and seeds. September to April. Dense terminal and axillary clusters of fleshy, spherical berries 0.2 to 0.3 inch (6 to 8 mm). Light green ripening to bright red in winter. Two hemispherical seeds.

Ecology. Occurs under forest canopies and near forest edges. Shade tolerant. Seedlings frequent in vicinity of old plantings. Varying leaf colors in the various cultivars, some of which do not produce viable seeds. Colonizes by root sprouts and spreads by animal-dispersed seeds.

History and use. Introduced from Eastern Asia and India in the early 1800s. Widely planted as an ornamental, now escaped and spreading from around old homes and recent landscape plantings. Sterile-seeded, reddish cultivars available.

Distribution. Found throughout the region with dense infestations in TN and LA. Mostly occurs in the north and west portions of the region and spreading southeast.

March

Shrubby Nonnative Lespedezas

Shrubby Lespedeza, *Lespedeza bicolor* Turcz. **LEBI2** **Fabaceae**
Other common name: bicolor
Thunberg's Lespedeza, *L. thunbergii* (DC.) Nakai **LETH4**

Plant. Perennial, erect, and much branched or ascending leguminous shrubs, 3 to 10 feet (1 to 3 m) in height. Stems clustered at the base with Thunberg's and single with bicolor. Trifoliate (3-leaflet) leaves, many small purple-to-white pea flowers, and single-seeded pods, from a woody root crown. Dormant brown plants remain upright most of the winter and may sprout at branches in the spring. Species probably hybridize to blur traits.

Stem. Arching branched, upright-to-ascending stems, 0.2 to 0.8 inch (0.5 to 2 cm) in diameter. Thunberg's often purple when young and bicolor often light tan to gray green. Appressed hairy to hairless.

Leaves. Alternate, 3-leaflet leaves. Thunberg's leaflets mostly narrowly elliptic to oblong, 0.8 to 2 inches (2 to 5 cm) long, and shrubby broadly elliptic to oval, 0.8 to 1.2 inches (2 to 3 cm) long, both with a hairlike tip. Lower surfaces lighter green than upper surfaces. Petioles 0.8 to 1.6 inches (2 to 4 cm) long. Stipules narrowly linear, 0.04 to 0.3 inch (1 to 8 mm) long.

Flowers. June to September. Clusters (racemes) 4 to 6 inches (10 to 15 cm) long growing from upper leaf axils—each cluster subtended by a tiny ovate bract—composed of 2 to 15 well-spaced, pealike flowers, more drooping in Thunberg's while shrubby are erect and extending beyond leaves. Each flower 0.3 to 0.6 inch (8 to 15 mm) long and beyond the upper leaves. Petals usually rosy purple in center and often grading to lighter shades, but can vary to white (many cultivars). Sepal teeth sharp with the lowest longer than the tube for Thunberg's while rounded and shorter than the tube for shrubby.

Fruit and seeds. August to March. Flat legume pod 0.2 to 0.3 inch (6 to 8 mm) long, broadly elliptic with pointed hairlike tip. Green becoming gray and densely appressed hairy, not splitting. Single seed 0.12 to 0.16 inch (3 to 4 mm) long, black for Thunberg's and mottled purple on green for shrubby.

Ecology. Planted widely in forest openings for wildlife food plots and soil stabilization to later encroach into adjoining stands. Reproduce and spread by abundant seed production even under a medium-to-dense overstory to exclude all other vegetation. Spread encouraged by burning that clears adjoining areas for seeding. Infestations seasonally changes from fire resistant in summer to hot burning in winter. Leguminous nitrogen fixers. Dense infestations hinder or stop recreational access, even by hunters.

History and use. Introduced from East Asia as ornamentals in the late 1800s. Later programs promoted uses for wildlife food and soil stabilization and improvement. Still sold even by States and planted for quail and whitetail deer food plots, while other birds and rodents do not eat the seed.

Distribution. Found throughout the region with scattered dense infestations in every State.

Shrubby Nonnative Lespedezas **6052**

Shrubby
August

Silverthorn, Thorny Olive

March

Ted Bodner

April

October

April

Elaeagnus pungens Thunb. **ELPU2** **Elaeagnaceae**
Other common names: spotted elaeagnus, thorny elaeagnus

Silverthorn, Thorny Olive **2037**

Plant. Evergreen, densely bushy shrub 3 to 25 feet (1 to 8 m) in height, with scattered sharp stubby branches and long limber projecting shoots that scramble up into adjacent trees. Thick leaves, silver-brown scaly beneath. Often found near ornamental and wildlife plantings as escaped single plants from animal-dispersed seeds.

Stem. Multiple stems and densely branched. Twigs brown, covered with brown scales, and hairy when young. Short shoots with small leaves become sharp-branched or unbranched thorns 0.4 to 1.6 inches (1 to 4 cm) long, and in second year produce leafy lateral branches, followed by flowers in fall. Lateral branches distinctly long, limber, and in late summer to spring extending beyond bushy crown and ascending into trees. Bark dark, drab and rough with projecting stubby thorns.

Leaves. Alternate, oval to elliptic and thick, 0.4 to 4 inches (1 to 10 cm) long and 0.2 to 2 inches (0.6 to 5 cm) wide. Irregular and wavy margins that may roll under. Blade surfaces silver scaly in spring becoming dark green or brownish green above and densely silver scaly with scattered brown scales beneath. Petioles 0.1 to 0.2 inch (4 to 5 mm) long, grooved above.

Flowers. October to December. Axillary clusters, each with 1 to 3 flowers, 0.4 inch (1 cm) long, silvery white to brown. Tubular with 4 lobes. Fragrant.

Fruit and seeds. March to June. Oblong, juicy drupe, 0.3 to 0.6 inch (1 to 1.5 cm) long, containing 1 nutlet. Whitish ripening to red and finely dotted with silvery to silvery-brown scales. Persistent shriveled calyx tube at tip.

Ecology. Fast-growing, weedy ornamental. Tolerant to shade, drought, and salt. Spreads by animal-dispersed seeds and occurs as scattered individuals, both in the open and under forest shade. Increases in size by prolific stem sprouts. Can climb into trees.

Resembles autumn olive (*E. umbellata* Thunb.) and Russian olive (*E. angustifolia* L.), both of which are deciduous and are further described in this book. Autumn olive has thin leaves with silver scales (not silver brown) and abundant reddish rounded berries in fall and early winter. Russian olive has silver scaly twigs and leaf surfaces, and many yellow fruit in fall and winter.

History and use. Introduced as an ornamental from China and Japan in 1830. Frequently planted for hedgerows and on highway right-of-ways and still used for landscaping. Mistakenly planted for wildlife to escape into nearby forests and pastures.

Distribution. Found as scattered plants or small infestations throughout the region.

October

Tropical Soda Apple

Solanum viarum **Dunal** **SOVI2** **Solanaceae**

Tropical Soda Apple **6095**

Plant. Upright, thorny perennial subshrub or shrub, 3 to 6 feet (1 to 2 m) in height, with leaves shaped like northern red oak leaves, clusters of tiny white flowers, and green-to-yellow golf-ball size fruit. Fruit sweet smelling and attractive to livestock and wildlife. Remains green over winter in most southern locations. **Caution: Fruit is poisonous.**

Stem. Upright to leaning, much branched, hairy, covered with broad-based white to yellow thorns.

Leaves. Alternate, 4 to 8 inches (10 to 20 cm) long and 2 to 6 inches (5 to 15 cm) wide. Margins deeply lobed (shaped like oak leaves). Velvety hairy with thorns projecting from veins and petioles. Dark green with whitish midveins above and lighter green with netted veins beneath.

Flowers. May to August (year-round in Florida). Terminal small clusters of 5-petaled white flowers. Petals first extended, then becoming recurved. Yellow to white fused stamens projecting from the center.

Fruit and seeds. June to November (year-round in Florida). Spherical, hairless, pulpy berry 1 to 1.5 inches (2.5 to 4 cm). Whitish then mottled green ripening to yellow. Each berry producing 200 to 400 reddish-brown seeds.

Ecology. Occurs on open to semishady sites. Viable seed in green or yellow fruit but not in white fruit. Reaches maturity from seed within 105 days. Persists by root crowns or green stems in warmer areas. Rapidly spreading by cattle and other livestock transportation and by wildlife-dispersed seeds as well as seed-contaminated hay, sod, and machinery.

Resembles horsenettle (*S. carolinense* L.), an 8- to 30-inch (20- to 80-cm) forb, which has similar but smaller fruit, long elliptic-to-ovate lobed leaves 3 to 5 inches (8 to 12 cm) long and 1 to 3 inches (2.5 to 8 cm) wide, and prickly yellow spines on stems and lower leaf veins but not on upper leaf. **Also resembles,** in stature and habitat, the closely related nonnative sticky nightshade (*S. sisymbriifolium*) which has deeply lobed lanceolate leaves and bright-red ripe fruit initially enclosed in a prickly husk.

History and use. Native to Argentina and Brazil and introduced into FL in the 1980s. No known use. A Federal listed noxious weed with an eradication program underway.

Distribution. Found in dense infestations in FL and the southern portions of MS, AL, GA, and SC, with outlying infestations in west NC and central TN. Mainly occurs in pastures and moving into the forest margins and openings.

May

Winged Burning Bush

Euonymus alatus (Thunb.) Siebold **EUAL8** **Celastraceae**
Other common names: winged wahoo, winged euonymus, burning bush

Winged Burning Bush **2042**

Plant. Deciduous, wing-stemmed, bushy shrub to 12 feet (4 m) in height, multiple stemmed and much branched. Canopy broad and leafy. Small obovate leaves green and turning bright scarlet to purplish red in fall. Paired purple fruit in fall on new growth.

Stem. Four corky wings or ridges appearing along young lime-green squarish twigs with wings becoming wider with age. Numerous opposite branches, with bases encircled by corky rings. Larger branches and bark becoming light gray.

Leaves. Opposite, obovate, and thin, 1 to 2 inches (2.5 to 5 cm) long and 0.4 to 0.8 inch (1 to 2 cm) wide. Tips tapering to an acute point. Margins finely crenate. Both surfaces smooth and hairless. Dark green with whitish midvein above and light green beneath, turning bright crimson to purplish red in fall. Petioles 0.04 to 0.16 inch (1 to 4 mm) long.

Flowers. April to May. Axillary pairs of small flowers at the ends of a Y-shaped 1-inch (2.5-cm) stem. Flowers inconspicuous, 0.2 to 0.3 inch (6 to 8 mm) across, greenish yellow, 5 lobed, pistil elongating as fruit forms.

Fruit and seeds. August to January. Dangling paired (or single) reddish capsules in leaf axils, turning purple and splitting in fall to reveal an orange fleshy-covered seed.

Ecology. Shade tolerant and invading forest understories, pastures, and coastal shrub lands. Colonizes by root suckers and spreads by animal-dispersed seeds.

Resembles the threatened and endangered native burning bush (*E. atropurpureus* Jacq.), which has erect-hairy lower leaf surfaces and petioles 0.3 to 0.8 inch (8 to 20 mm) long. **Also resembles** the larger-leaved species of blueberry (*Vaccinium* spp.), but their leaves are alternate. **Possibly resembles** rusty blackhaw (*Viburnum rufidulum* Raf.), which also has opposite leaves but distinguished by their larger size and leathery texture. Dormant twigs **may resemble** winged elm (*Ulmus alata* Michx.) and sweetgum (*Liquidambar styraciflua* L.), which are usually 2 winged instead of 4 winged.

History and use. Introduced from Northeast Asia in the 1860s. Widely planted as an ornamental and for highway beautification.

Distribution. Found throughout the region except AR, TX, and OK with dense infestations in KY and VA.

April

Amur Peppervine, Porcelain Berry

Ampelopsis brevipedunculata (Maxim.) Trautv. **AMCO2** **Vitaceae**

Amur Peppervine, Porcelain Berry

Plant. Deciduous, woody vine of the grape family to 20 feet (5 m) long, running and climbing over shrub and tree crowns and rock faces by clinging tendrils, forming thicket and arbor infestations. Heart-shaped leaves with up to 6 deep lobes and toothed margins, axillary flat clusters of inconspicuous yellowish flowers in spring, and multicolored spherical fruit of white, green, or blue in the fall and winter.

Stem. Woody vine to 4 inches (10 cm) diameter, climbing by tendrils with forked ends and fine hairs (tendrils not forked on native grapes). Tendrils occur on new growth, opposite leaves. New stems are whitish green, smooth to lightly hairy, slightly square with regularly spaced swollen nodes, increasingly with raised dots (lenticels) that become corky and reddish, eventually forming linear gray-reddish barky patches. Bark glossy light gray becoming gray and rough with persistent swollen nodes. Pith is white, while native grape pith is brown.

Leaves. Alternate, simple and heart shaped in outline but variable in form from entire with 1 to 3 tips to those with multiple incised, rounded to scalloped lobes between 3 to 5 prominent whitish veins radiating from the base. Margins coarsely toothed with distinctive fine whitish hair tips. Dark green to blue green, shiny above and fine hairy beneath, becoming yellow in fall. New leaves tiny at branch tips, progressively increasing up to 6 inches (15 cm) long and 4 inches (10 cm) wide. Petioles 2.5 inches (6 cm) long, light green and hairless, with greatly swollen bases.

Flowers. June to August. Numerous tiny-branched clusters (cymes), with up to 40 flowers, opposite new leaves and fruit in midsummer. Flowers tiny with 5 spreading white petals (grape petals touch at tips) and 5 yellow erect stamens (hand lens may be required).

Fruit and seeds. July to January. Drupes in clusters, shiny, spherical, to 0.5 inch (1.2 cm) wide, green tipped with a persistent pistil and turning whitish, yellow, purple, turquoise, and porcelain blue (thus the common name), with all colors sometimes present. Each drupe contains 2 to 4 seeds. Persist in winter at most leaf axils.

Ecology. Occurs on a wide range of sites and grows rapidly to form exclusive infestations along forest edges. Found as scattered plants to extensive infestations in forest openings, margins, and roadsides as well as along stream margins and riverbanks. Areas from full sun to partial shade. Colonizes by prolific vine growth that roots at nodes. Seeds spread by birds and other animals, may be viable in the soil for several years.

Resembles grape vines (*Vitis* spp.) but can be distinguished by the whitish pith versus grape's darker pith, and forked tendrils versus grape's linear tendrils. **Also resembles** the native heartleaf peppervine (*A. cordata* Michx.) with unlobed leaves and hairless stems.

History and use. Introduced from Northeast Asia in the 1800s as a landscape plant for the uniquely colored berries. Varieties are still sold in the plant trade.

Distribution. Found throughout the region except AR, TX, and OK with frequent dense infestations in KY, VA, TN, and NC.

Steven T. Manning

Climbing Yams

Air Yam July

Chinese Yam November

Chinese Yam July

Air Yam July

Air Yam December

Water Yam August

Fred Nation

Air Yam December

Air yam, *Dioscorea bulbifera* L. DIBU **Dioscoreaceae**
Other common name: air potato
Chinese yam, *D. oppositifolia* L. DIOP
Synonym: *D. batatas* Dcne.
Other common name: cinnamon vine
Water yam, *D. alata* L. DIAL2
Other common names: winged yam, white yam

Plant. Herbaceous, high climbing vines to 65 feet (20 m) long, infestations covering shrubs and trees. Twining and sprawling stems with long-petioled heart-shaped leaves. Spreading by dangling potato-like tubers (bulbils) at leaf axils and underground tubers. Monocots.

Stem. Twining and covering vegetation, branching, hairless. Internode cross sections round for air yam to angled for Chinese and water yams. Water yam nodes winged and reddish. All stems dying back in winter leaving some bulbils attached.

Leaves. Alternate (air) or combination of alternate and opposite (Chinese and water). Heart shaped to triangular with elongated tips, thin and hairless, 4 to 8 inches (10 to 20 cm) long and 2 to 6 inches (5 to 15 cm) wide. Long petioled. Basal lobes broadly rounded (air) or often angled (Chinese and water). Margins smooth. Veins curved and converging at tip and base. Dark green with slightly indented curved veins above (quilted appearance) and lighter green beneath. Chinese yam leaves turning bright yellow in fall.

Flowers. May to August. Rare, small, panicles or spikes to 4.5 inches (11 cm) long in axils male and female flowers on separate plants (dioecious). Green to white. Fragrant, with Chinese yam having a cinnamon fragrance (thus the common name cinnamon vine).

Fruit and seeds. June to September (and year round). One to 4 aerial tubers (bulbils) resembling miniature potatoes or yams occur at leaf axils, eventually dropping and sprouting to form new plants. Shape spherical to ovoid (air and Chinese) to oblong (water). Texture smooth with dimples (air) to warty (Chinese) to rough (water). Air yam to 5 inches (12 cm) long, Chinese yam to 1 inch (2.5 cm) long, and water yam to 4 inches (10 cm) long and 1.2 inches (3 cm) wide. Very rarely have capsules and winged seeds, which have questionable viability.

Ecology. Rapid growing and occurring on open sites: water yams in FL, air yams extending from FL to adjacent States, and Chinese yams in all States except FL. All dying back during winter but able to cover small trees in a year, with old vines providing trellises for regrowth. Spread and persist by underground tubers and abundant production of aerial tubers (bulbils), which drop and form new plants and can spread by water. Chinese yam aerial tubers persist only 1 year.

Resemble greenbriers (*Smilax* spp.) with most having thorns and/or green-to-purple berries but no aerial potatoes. **Also resemble** several native *Dioscorea* species that do not form dense vine infestations nor have aerial tubers (bulbils): wild yam (*D. villosa* L.) with hairy upper leaf surfaces; native Florida yam (*D. floridana* Bartlett); and, only in Florida, nonnative Zanzibar yam (*D. sansibarensis* Pax.). Chinese yam leaves are a similar shape to the native vine, Carolina coralbead (*Cocculus carolinus* (L.) DC.), and morning-glory (*Ipomea* spp.) while their veins are not curved tip to base.

History and use. Introduced from Africa (air) and Asia (Chinese and water) as possible food sources in the 1800s. Ornamentals often spread by unsuspecting gardeners intrigued by the dangling yams. Presently cultivated for medicinal use.

Distribution. Chinese yam found scattered throughout the region, while air and water yam occur along the southern portions of the Gulf Coast States and throughout FL.

Climbing Yams **3030**

Air Yam October

Five-Leaf Akebia, Chocolate Vine

Steven T. Manning

Steven T. Manning

Nancy Fraley

Nancy Loewenstein

Akebia quinata (Houtt.) Decne. **AKQU** **Lardizabalaceae**

Five-Leaf Akebia, Chocolate Vine

Plant. Woody, semi-evergreen or evergreen vine to 40 feet (12 m) long, climbing by twining to dangle and sprawl in tree and shrub crowns and/or forming solid ground cover, up to 1 foot (30 cm) deep. Small, dark green, palmately compound leaves with 5 elliptical leaflets on long petioles. Infrequent, showy, dangling purple flowers appear with leaves in spring with female flowers infrequently yielding sausage-shaped pods in fall.

Stem. Woody, round, to 4 inches (10 cm) in diameter at the base, with numerous branching stems twining for support on plants or natural trellises or forming ground cover and rooting where nodes contact soil. Lime green and smooth, becoming dotted with many brownish dots (lenticels), then light and dark gray striated and finally light gray speckled with raised gray dots. Leaf scars circular and cleft, protruding alternately or in clusters along the stem.

Leaves. Alternate, palmately compound, usually 5 leaflets, obovate to elliptic or oblong, 1 to 3 inches (2.5 to 8 cm) long and 0.4 to 1.5 inches (1 to 4 cm) wide, terminal leaflet usually the largest. New leaves purple tinged turning dull blue green, midvein and 2 lateral veins lighter above and leaflet whitish beneath. Margins entire, tip notched or blunt with tiny hair. Rachis (leaf stem) to 4 inches (10 cm) long, while leaflet stems (petiolules) shorter to only 1 inch (2.5 cm) long. Where deciduous, leaves appear early in spring with no color change before dropping in winter at the first frost.

Flowers. March to April. Small purple-to-violet flowers of 3 sepals (no true petals) in long-stalked clusters of 2 to 5 that appear with leaves and can be unnoticed within the foliage. Fragrance likened to chocolate. Male flowers terminal in clusters, smaller, 0.5 to 0.75 inch (1.2 to 1.6 cm) wide and female flowers, 1 to 1.5 inches (2.5 to 4 cm) wide, extend outward on long stalks.

Fruit and seeds. May to ripening in October. Rarely fruiting. Dangling clusters of 1 to 5 sausage-shaped, fleshy pods, 2 to 4 inches (5 to 10 cm) long, tipped like a banana and having a lengthwise suture along one side. Light green, turning pink to purplish with lighter speckles and a waxy coating. When ripe the skin splits to reveal a pulpy, edible inner core that splits further to expose many (100+) imbedded black seeds.

Ecology. Occurs on a wide range of sites, somewhat drought tolerant but prefers moist soils, full sun to partial shade. The dense ground cover displaces native plants and wildlife while vigorous climbing vines cover and kill small trees and shrubs. Colonizes by prolific vine growth that root and sprout. Infrequent seeds spread by birds and other animals.

Resembles bottlebrush buckeye (*Aesculus parviflora* Walt.) when young due to similar palmately compound leaves with 5 leaflets, but buckeye leaves have pointed tips and pronounced lateral veins.

History and use. Introduced from Eastern Asia in 1845 as an ornamental. Still sold in the plant trade with several varieties that produce fertile seeds when cross-pollinated.

Distribution. Found in scattered infestations throughout the region but not yet reported in LA, TX, and OK.

Nancy Loewenstein

Japanese Honeysuckle

Ted Bodner

Ted Bodner

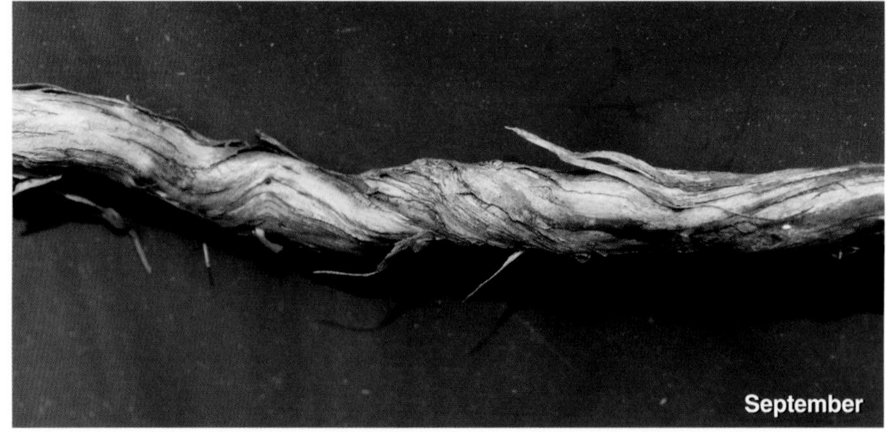

Lonicera japonica Thunb. **LOJA** **Caprifoliaceae**

Japanese Honeysuckle **3101**

Plant. Semi-evergreen to evergreen woody vine, high climbing and trailing to 80 feet (24 m) long, branching and often forming spotty to extensive arbors in lower and upper forest canopies and/or ground cover under canopies and in new forests, rooting at nodes along leaf-covered vines (stolons).

Stem. Slender woody vine becoming stout to 2 inches (5 cm) in diameter, with cross-section round and opposite branching. Brown and hairy becoming tan barked, fissured, and sloughing with age. Rooting at low nodes.

Leaves. Opposite, broadly ovate to elliptic to oblong, base rounded, tips blunt pointed to round. Length 1.6 to 2.6 inches (4 to 6.5 cm) and width 0.8 to 1.5 inches (2 to 4 cm). Margins entire but often lobed in early spring. Both surfaces smooth to rough hairy, with undersurface appearing whitish.

Flowers. April to August. Axillary pairs, each 0.8 to 1.2 inches (2 to 3 cm) long, on a bracted stalk. White (or pink) and pale yellow. Fragrant. Thin tubular, flaring into 5 lobes in 2 lips (upper lip 4 lobed and lower lip single lobed), with the longest lobes roughly equal to the tube. Five stamens and 1 pistil, all projecting outward and becoming curved. Persistent sepals.

Fruit and seeds. June to March. Nearly spherical, green ripening to black, glossy berry 0.2 inch (5 to 6 mm) on stalks 0.4 to 1.2 inches (1 to 3 cm) long. Two to 3 seeds.

Ecology. Most commonly occurring invasive plant in the South, overwhelming and replacing native flora in all forest types over a wide range of sites or occurring as scattered plants. Often coexisting with other invasive plants. Occurs as dense infestations along forest margins and right-of-ways as well as under dense canopies and as arbors high in canopies. Shade tolerant. Persists by large woody rootstocks and spreads mainly by rooting at vine nodes and less by animal-dispersed seeds. Infrequently seeding within forest stands with very low germination. Seed survival in the soil is less than 2 years.

Resembles yellow jessamine [*Gelsemium sempervirens* (L.) W.T. Aiton], which has narrower leaves and hairless stems. **Also resembles** native honeysuckles (*Lonicera* spp.) that usually have reddish hairless stems and hairless leaves and do not form extensive infestations.

History and use. Introduced from Japan through England in the early 1800s. Traditional ornamental, valued as deer browse, with some value for erosion control. Still planted in wildlife food plots to encroach on adjacent lands.

Distribution. The most pervasive invasive plant throughout the region with the most frequent and dense infestations in east-central AL and a sizeable number in central MS and TN, as well as northwest SC.

April

Kudzu

June

October

Ted Bodner

April

July

August

November

Ted Bodner

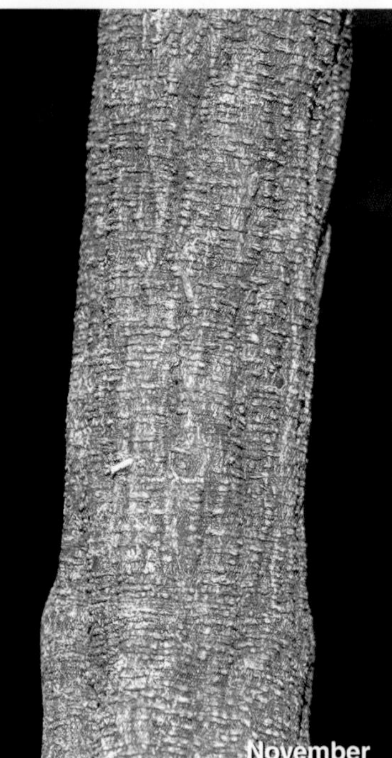
November

Pueraria montana **(Lour.) Merr. PUMOL** **Fabaceae**
Synonyms: *P. lobata* (Willd.) Ohwi, *P. montana* var. *lobata* (Willd.) Maesen & S. Almeida

Plant. Deciduous twining, trailing, deep mat-forming, ropelike woody leguminous vine, 35 to 100 feet (10 to 30 m) long with 3-leaflet leaves. Large semiwoody tuberous roots reaching depths of 3 to 16 feet (1 to 5 m) with a knot- or ball-like root crown on top at the soil surface where vines originate. Leaves and small vines dying with first frost and matted dead leaves persistent during winter.

Stem. Woody vines to 10 inches (25 cm) in diameter, round in cross section, with infrequent branching. Stems succulent and yellow green with dense, erect golden hairs and upward-matted silver hairs, aging to ropelike and light gray barked. Frequent unswollen nodes root when on the ground to form new plants when interconnecting vines die, eventually growing large with age to form root crowns 1 to 10 inches (2.5 to 25 cm) wide. Mature bark eventually rough, rigid, and usually dark brown.

Leaves. Alternate, pinnately compound 3-leaflet leaves, each leaflet 3 to 7 inches (8 to 18 cm) long and 2.5 to 8 inches (6 to 20 cm) wide. Usually slightly lobed (unless in shade) with a 2-lobed symmetric middle leaflet and two 1-lobed side leaflets. Tips pointed. Margins thin membranous and fine golden hairy. Leaflet stems (petiolules) swollen near leaftlets. Petrioles 6 to 12 inches (15 to 30 cm) long, long hairy, base swollen, with deciduous stipules.

Flowers. June to September. Axillary slender clusters (racemes), 2 to 12 inches (5 to 30 cm) long, of pealike flowers in pairs (or threes) from raised nodes spiraling up the stalk, opening from the base to top. Petals lavender to wine colored with yellow centers. Fragrant, often reported with the scent of grapes.

Fruit and seeds. September to January. Clustered dry, flattened legume pods (bulging above the seeds) each 1.2 to 3 inches (3 to 8 cm) long and 0.3 to 0.5 inch (8 to 12 mm) wide. Green ripening to tan with stiff golden-brown hairs. Falling whole or splitting on 1 to 2 sides to release a few ovoid seeds.

Ecology. Can grow 1 foot (30 cm) per day in spring and 60 feet (18 m) per year. Occurs in old infestations, along right-of-ways, forest edges, and stream banks. Forms dense mats over the ground, debris, shrubs, and mature trees forming dense patches by twining on objects less than 4 inches (10 cm) in diameter. Colonizes by vines rooting at nodes and spreads by wind-, animal-, and water-dispersed seeds. Seed viability variable by habitat and across the region. Leguminous nitrogen fixer.

History and use. Introduced from Japan and China in early 1900s with continued seed importation. Limited use for erosion control, livestock feed, and folk art.

Distribution. Found throughout the region with scattered dense infestations in every State. Especially frequent in MS, AL, GA, and northwest SC.

Kudzu **3123**

July

Erwin Chambliss

Nonnative Ivies

English Ivy shown in all images

English Ivy, *Hedera helix* L. HEHE **Araliaceae**
Atlantic Ivy or Irish Ivy, *H. hibernica* (G. Kirchn.) Bean HEHI12
Colchis Ivy or Persian Ivy, *H. colchica* (K. Koch) K. Koch HECO20

Plant. Evergreen woody vines climbing to 90 feet (28 m) by clinging aerial roots and trailing to form dense ground cover. Thick dark-green leaves with whitish veins and 3 to 5 pointed lobes when juvenile, colchis ivy with 3 slight tips and few sharp teeth. Maturing at about 10 years into erect plants or branches with unlobed leaves and terminal flower clusters that yield blackish-to-purplish berries. Hundreds of cultivars vary in leaf size and color. **Caution: Fruit toxic to humans, and plant contact triggers dermatitis in sensitive individuals.**

Stem. Woody slender vines when a ground cover and growing to 10 inches (25 cm) in diameter when climbing infested trees and rocks by many fine to stout aerial rootlets. Vines pale green (sometimes reddish tinged), rooting at nodes, becoming covered with gray-brown shiny bark, segmented by encircling and raised leaf scars, and roughened by tiny ridges. Bark light gray to brown, bumpy and gnarly, with aerial rootlets that exude a gluelike substance to tightly cling to vertical structures. Older vines sometimes grown together where crossed.

Leaves. Alternate, with shapes varying according to age—typical juvenile plants having 3 to 5 pointed lobes and mature plants broadly lanceolate (English and Atlantic) to cordate (colchis). English generally less than 3.3 inches (8 cm) wide, Atlantic up to 4 inches (10 cm) wide, and colchis about 4 or more inches (10 cm) wide. All thick and waxy, dark green (some whitish variegated cultivars) with light-green veins radiating from the petiole and pale green beneath. Colchis dull green above, drooping, and fragrant when crushed. Hairs whitish on English, yellowish brown to rusty brown on Atlantic, and scalelike on colchis (requires a hand lens). Petioles to 6 inches (15 cm) long and pale green.

Flowers. June to October. Terminal hairy-stemmed umbel clusters of small greenish-yellow flowers on mature plants. Five thick and pointed petals, 0.1 inch (3 mm) long. Each petal radiating from a 5-sided domed green floral disk, 0.1 inch (3 mm) wide, tipped by a short pistil.

Fruit and seeds. October to May. Clusters of spherical drupes, 0.2 to 0.3 inch (7 to 8 mm). Pale green in late summer ripening to dark blue to black in late winter to spring.

Ecology. Thrive in moist open forests, but adaptable to a range of moisture and soil conditions, including rocky cliffs. Avoid wet areas. Shade tolerance allowing early growth under dense stands, but becoming adapted to higher light levels with maturity. Grow very aggressively once established. Amass on infested trees, decreasing vigor, and increasing chance of windthrow. Increased sunlight promotes flowering and fruiting. English serves as a reservoir for bacterial leaf scorch that infects oaks (*Quercus* spp.), elms (*Ulmus* spp.), and maples (*Acer* spp.). Spread by bird-dispersed seeds and colonization by trailing and climbing vines that root at nodes. Drupes mildly toxic, discouraging overconsumption by birds. Foliage consumed by some butterfly larvae. English rarely produces fertile seeds on the Gulf Coast.

Resemble grape (*Vitis* spp.) which has a leaf that is similarly shaped but not thick and often hairy.

History and use. Introduced from England, Europe, and Asia in colonial times. Traditional ornamentals and still widely planted. English a source of varnish resin, dye, and tanning substances.

Distribution. English ivy found throughout the region with scattered dense infestations in every State. Especially frequent in urban forests. Currently Atlantic and colchis ivies only in NC and SC.

Nonnative Ivies **3071**

September

Nonnative Wisterias

Chinese wisteria, *Wisteria sinensis* (Sims) DC. **WISI** **Fabaceae**
Japanese wisteria, *W. floribunda* (Willd.) DC. **WIFL**

Plant. Deciduous high climbing, twining, or trailing leguminous woody vines (or cultured as shrubs) to 70 feet (20 m) long. Chinese and Japanese wisteria difficult to distinguish due to hybridization.

Stem. Woody vines to 10 inches (25 cm) in diameter with infrequent alternate branching. Twigs densely short hairy. Older bark of Chinese wisteria tight and dark gray with light dots (lenticels) compared to white bark of Japanese wisteria.

Leaves. Alternate, odd-pinnately compound 4 to 16 inches (10 to 40 cm) long, with 7 to 13 leaflets (Chinese) or 13 to 19 leaflets (Japanese), and stalks with swollen bases. Leaflets oval to elliptic with tapering pointed tips, 1.6 to 3 inches (4 to 8 cm) long and 1 to 1.4 inches (2.5 to 3.5 cm) wide. Hairless to short hairy at maturity but densely silky hairy when young. Margins entire and wavy. Sessile or short petioled.

Flowers. March to May. Dangling and showy, stalked clusters (racemes) appearing when leaves emerge, 4 to 20 inches (10 to 50 cm) long and 3 to 3.5 inches (7 to 9 cm) wide. All blooming at about the same time (Chinese) or gradually from base (Japanese). Pealike flowers, corolla lavender to violet (to pink to white). Fragrant.

Fruit and seeds. July to November. Flattened legume pod, irregularly oblong to oblanceolate, 2.5 to 6 inches (6 to 15 cm) long and 0.8 to 1.2 inches (2 to 3 cm) wide. Velvety hairy, greenish brown to golden, splitting on 2 sides to release 1 to 8 flat, round, brown seeds, each 0.5 to 1 inch (1.2 to 2.5 cm) in diameter.

Ecology. Form dense infestations where previously planted. Occur on wet to dry sites. Colonize by vines twining and covering shrubs and trees and by runners rooting at nodes when vines covered by leaf litter. Seeds water-dispersed along riparian areas. Large seed size a deterrent to animal dispersal.

Resemble native or naturalized American wisteria [*W. frutescens* (L.) Poir.], which occurs in wet forests and edges and sometimes forms large entanglements, flowers in June to August after leaves develop, and has 6-inch (15-cm) flower clusters, 9 to 15 leaflets, thin hairless, papery pods, and slender old vines. **Also may resemble** trumpet creeper [*Campsis radicans* (L.) Seem. ex Bureau], which has leaflets with coarsely toothed margins and white-hairy prominent veins beneath.

History and use. Introduced from Asia in the early 1800s. Traditional southern porch vines and still planted by mistaken gardeners.

Distribution. Found throughout the region with scattered dense infestations in every State. Especially frequent in SC and southwest AL.

Nonnative Wisterias **3251**

Chinese wisteria April

Oriental Bittersweet

December

May

May

October

October

August

December

November

October

October

Celastrus orbiculatus **Thunb. CEOR7** **Celastraceae**
Other common name: Asian bittersweet

Oriental Bittersweet **3026**

Plant. Deciduous, twining and climbing woody vine to 60 feet (20 m) in tree crowns, forming thickets and arbor infestations. Elliptic to round-tipped leaves, axillary dangling clusters of inconspicuous yellowish flowers in spring, and green spherical fruit that split to reveal 3-parted showy scarlet berries in winter.

Stem. Woody vine to 4 inches (10 cm) diameter, twining and arbor forming, with many alternate drooping branches growing at angles and eventually becoming straight. Vigorous twigs with sharp bud scale tips. Reddish brown with many raised whitish corky dots (lenticels), often angular or ridged, becoming tan to gray. Branch scars of fruit clusters semicircular, each with a tiny corky shelf projection. Bark dark grayish brown with irregular netted ridges.

Leaves. Alternate, 1.2 to 5 inches (3 to 12 cm) long. Variable shaped, long tapering tipped when young becoming larger and round tipped when mature. Margins finely blunt toothed. Dark green becoming bright yellow in late summer to fall. Base tapering into 0.4- to 1.2-inch (1- to 3-cm) petiole.

Flowers. May. Numerous tiny-branched axillary clusters (cymes), each with 3 to 7 inconspicuous orange-yellow flowers. Five petals. Male and female flowers can occur on the same or different plants.

Fruit and seeds. August to January. Dangling clusters of spherical 0.5-inch (1.2 cm) capsules, tipped with a persistent pistil. Green turning yellow orange then tan. In autumn, splitting and folding upward to reveal 3 fleshy scarlet sections, each containing 2 white seeds. Persistent in winter at most leaf axils.

Ecology. Occurs on a wide range of sites mainly along forest edges. Found as scattered plants to extensive infestations in forest openings, margins, and roadsides as well as in meadows. Shade tolerant with high seed germination under canopies. Colonizes by prolific vines that root at nodes, and seedlings from prolific seed spread mainly by birds, possibly other animals and humans collecting and discarding decorative fruit-bearing vines.

Resembles American bittersweet (*C. scandens* L.), which has only terminal flowers and fruit, leaves usually twice as large but absent among the flowers and fruit, grayish and nonridged twigs, and blunt bud-scale tips. Hybridization occurs between the 2 species. **Also resembles** grape vines (*Vitis* spp.) in winter but can be distinguished by persistent scarlet fruit versus grapes.

History and use. Introduced from Asia in 1736. Very showy ornamental with berried vines that are traditionally collected as home decorations in winter, which promotes spread when inappropriately discarded.

Distribution. Found throughout the region except FL, TX, and OK with frequent and dense infestations in east KY, west NC, and north VA.

October

Vincas, Periwinkles

Bigleaf periwinkle April

Common periwinkle leaves Bigleaf periwinkle leaves May

Bigleaf periwinkle April

Bigleaf periwinkle March

Nancy Loewenstein

Common periwinkle flower Bigleaf periwinkle flower Summer

Barry Rice

Common periwinkle, *Vinca minor* L. VIMI2 **Apocynaceae**
Bigleaf periwinkle, *V. major* L. VIMA

Plant. Evergreen to semi-evergreen vines, somewhat woody, trailing or scrambling to 3 feet (1 m) long and upright to 1 foot (30 cm). Violet pinwheel-shaped flowers.

Stem. Slender (common periwinkle) to stout (bigleaf periwinkle), succulent becoming somewhat woody (tough to break) with flowering branches erect and jointed at axils. Hairless and smooth. Dark green at base to light green upward with a reddish tinge.

Leaves. Opposite. Glossy and hairless, somewhat thick, with margins slightly rolled under. Common periwinkle narrow elliptic, 0.8 to 1.8 inches (2 to 4.5 cm) long and 0.4 to 1 inch (1 to 2.5 cm) wide, with petioles 0.1 inch (1 to 3 mm) long. Bigleaf periwinkle heart shaped to somewhat triangular to elliptic, 1.5 to 2.5 inches (4 to 6 cm) long and 1 to 1.5 inches (2.5 to 4 cm) wide, with petioles 0.2 to 0.4 inch (5 to 10 mm) long. Blades dark green with whitish lateral- and mid-veins above and lighter green with whitish midveins beneath. Some varieties variegated.

Flowers. March to May (sporadically May to September). Axillary, usually solitary. Violet to blue lavender (to white), with 5 petals radiating pinwheel-like at right angles from the floral tube. Common periwinkle 1 inch (2.5 cm) wide with a 0.3- to 0.5-inch (8- to 12-mm) long tube. Bigleaf periwinkle 1.5 to 2 inches (4 to 5 cm) wide with a 0.6- to 0.8-inch (1.5- to 2-cm) long tube. Five slender lanceolate sepals, about 0.4 inch (1 cm) long, hairy margined.

Fruit and seeds. May to July. Slender, cylindrical fruit to 2 inches (5 cm) long. Becoming dry and splitting to release 3 to 5 infertile seeds.

Ecology. Found around old homesite plantings and scattered in open to dense canopied forests. Form mats and extensive infestations, even under forest canopies, by vines rooting at nodes.

Resemble partridgeberry (*Mitchella repens* L.), which has cordate leaves, twin white flowers, and red berries. **Also may resemble** yellow jessamine [*Gelsemium sempervirens* (L.) W.T. Aiton], which has wider spaced leaves and reddish stems, often white waxy. **Also resembles** winter creeper [*Euonymus fortunei* (Turcz.) Hand.-Maz.], which has stout light-green vines, leathery leaves, and no showy flowers.

History and use. Introduced from Europe in 1700s. Ornamental ground cover, commonly sold and planted by gardeners.

Distribution. Found throughout the region with scattered dense infestations in every State.

Vincas, Periwinkles **3211**

Bigleaf periwinkle April

Winter Creeper

Euonymus fortunei (Tursz.) Hand.-Maz. **EUFO5** **Celastraceae**
Synonym: *E. hederaceus* Champ. ex Benth.
Other common names: climbing euonymus, gaiety

Winter Creeper **3042**

Plant. Evergreen woody vine climbing to 70 feet (22 m) and clinging by aerial roots or rooting at nodes, or standing as a shrub to 3 feet (1 m) in height. Leaves thick and dark green or green- or gold-white variegated on green stems. The nonflowering juvenile climbing phase, upon reaching high enough into the crowns of trees, develops into a flowering phase that does not have climbing rootlets. Pinkish-to-red capsules splitting open in fall to expose fleshy orange seeds.

Stem. Twigs stout, lime green, and hairless becoming increasingly dusted and streaked with light-gray reddish corky bark. Patches or lines of protruding aerial roots grow on lower surfaces or where touching supporting structures. Branches opposite, leaf scars thin upturned white crescents, and branch scars jutting and containing a light semicircle. Older stems covered with gray corky bark becoming fissured and then checked.

Leaves. Opposite, broadly oval, moderately thick, with bases tapering to petiole. One to 2.5 inches (2.5 to 6 cm) long and 1 to 1.8 inches (2.5 to 4.5 cm) wide. Margins finely crenate, somewhat turned under, to wavy. Blades smooth glossy, hairless, dark green with whitish veins above and light green beneath. Some varieties variegated, with white or golden margins. Petioles 0.15 to 0.4 inch (0.4 to 1 cm) long.

Flowers. May to July. Axillary clusters of small greenish-yellow inconspicuous flowers at the ends of Y-shaped stems, each flower 0.1 inch (2 to 3 mm) wide. Five petals. Pistils soon elongating with fruit.

Fruit and seeds. September to November. Dangling paired or single pinkish-to-red capsules, 0.2 to 0.4 inch (5 to 10 mm) long, splitting to reveal 4 tightly clustered seeds with orange-to-red fleshy coats.

Ecology. Forms dense ground cover and can climb trees, eventually overtopping them. Climbing vines produce fruit. Cold and shade tolerant, occurring under dense stands but avoiding wet areas. Colonizes by trailing and climbing vines that root at nodes, and fleshy-coated seeds spread by birds, other animals, and water.

Resembles the larger-leaved species of blueberry (*Vaccinium* spp.) but their leaves are alternate. **Also resembles** native partridgeberry (*Mitchella repens* L.), a creeping vine with opposite oval or cordate leaves less than 1 (2.5 cm) inch long and wide, white twin flowers and red berries; and the nonnative vincas (*Vinca* spp.), trailing vines with similar opposite leaves but margins are rolled under and flowers violet-to-blue pinwheels.

History and use. Introduced from Asia in 1907. Ornamental ground cover.

Distribution. Found throughout the region except FL, LA, TX, and OK with scattered dense infestations in every other State. Especially frequent in MS, AL, NC, KY, and central VA.

October

Bamboos

Golden bamboo, *Phyllostachys aurea* **PHAU8** **Poaceae**
Carr. ex A.& C. Rivière
Other invasive bamboos: *Phyllostachys* spp. and *Bambusa* spp.

Plant. Perennial infestation-forming bamboos, 16 to 40 feet (5 to 12 m) in height, with jointed cane stems and whorls of branches at each node. Bushy tops of lanceolate leaves along branches, in fan clusters, often golden green. Plants arising from branched rhizomes.

Stem. Solid jointed canes 1 to 6 inches (2.5 to 15 cm) in diameter. Hollow between joints. Golden to green to black. Branches wiry and projecting from joints. Lower shoots and branches with loose papery sheaths that cover the ground when shed.

Leaves. Alternate, grasslike, often in fan clusters. Blades long and lanceolate, 3 to 10 inches (8 to 25 cm) long and 0.5 to 1.5 inches (1.3 to 4 cm) wide. Often golden, sometimes green or variegated. Veins parallel. Hairless except for large hairs at base of petiole, which shed with age. Deciduous sheaths encasing stem.

Flowers. Flowers very rarely.

Seeds. Seeds very rarely.

Ecology. Common around old homesites and now escaped. Colonize by rhizomes and less so by stolons with infestations rapidly expanding after disturbance. General dieback after period flowering and seeding (about every 7 to 12 years) resulting in standing dead canes and new shoots.

Resemble switchcane [*Arundinaria gigantea* (Walter) Muhl. and other native *Arundinaria* spp.], the only native bamboolike canes in the South, distinguished by its persistent sheaths on the stem, short alternate branches or branch clusters, and its lower height —usually only 6 to 8 feet (2 to 2.5 m). **Also resemble** giant reed (*Arundo donax* L.), also described in this book.

History and use. All native to Asia. Widely planted as ornamentals and for fishing poles.

Distribution. Found throughout the region with scattered dense infestations in every State.

Bamboos **4130**

Chinese Silvergrass

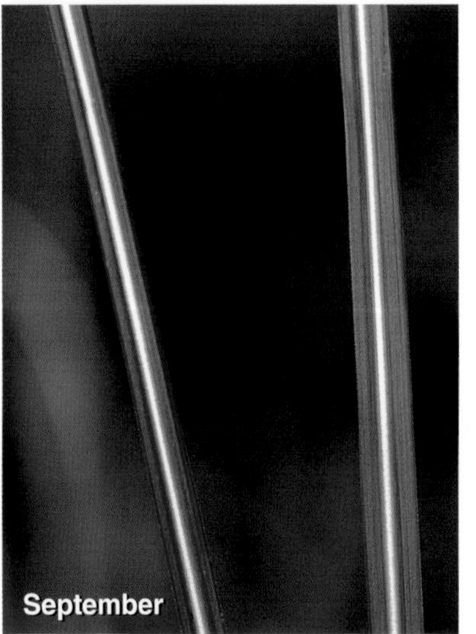

Miscanthus sinensis Andersson **MISI** **Poaceae**

Chinese Silvergrass **4085**

Plant. Tall, densely bunched, perennial grass, 5 to 10 feet (1.5 to 3 m) in height. Long-slender upright-to-arching leaves with whitish upper midveins. Many loosely plumed panicles in late summer turning silvery to pinkish in fall. Dried grass standing with some seed heads during winter. Seed viability variable depending on cultivar.

Stem (culm). Upright-to-arching, unbranched originating in tufts from base. Covered with overlapping leaf sheaths until stem appears with flower plume in late summer.

Leaves. Alternate, long linear, upright-to-arching (persisting and curly tipped when dried) to 40 inches (1 m) long and less than 0.8 inch (2 cm) wide. Blades green to variegated (light green striped) with whitish collars. Midvein white above and green ridged beneath. Tufted hairs at throat, sheath margins, and ligule, but otherwise hairless. Margins rough.

Flowers. August to November. Terminal, plumed panicles, much branched and drooping, 4 to 15 inches (10 to 38 cm) long and 2 to 8 inches (5 to 20 cm) wide. Silvery to pinkish, showiest in fall. Stalk appressed-rough hairy.

Seeds. September to January. Grain hidden, husks membranous, yellowish brown to slightly reddish, sparsely hairy, with twisted tip.

Ecology. Forms extensive infestations by escaping from older ornamental plantings to roadsides, forest margins, right-of-ways and adjacent disturbed sites, especially after burning. Shade tolerant. Highly flammable and a fire hazard.

Resembles giant reed (*Arundo donax* L.), which has wide cornlike leaves jutting from a central stalk and is described in this book. **Also resembles** Uruguayan pampas grass [*Cortaderia selloana* (Schult. & Schult. f.) Asch. & Graebn.], another nonative invasive grass and also commonly planted as an ornamental, which is usually larger in size with a tighter plume branched from a central stalk. **Also resembles** common reed [*Phragmites australis* (Cav.) Trin. ex Steud.], which has a similar large hairy seed head, but fanned in a loose plume and not erect, and which occurs mainly near swamps, marshes and other wet habitats. **Also resembles** the native sugarcane plumegrass [*Saccharum giganteum* (Walter) Pers.] but its leaves are mainly on the lower stem and do not have a distinctive white midvein.

History and use. Introduced from Eastern Asia. Still widely sold and increasingly planted as an ornamental. Several varieties imported and sold. Cultivars vary widely in fertility and percent of seed viability.

Distribution. Found in scattered dense infestations in NC, west KY, south VA, east TN, and south GA with occasional infestations elsewhere throughout the region except TX, OK, and AR.

October

Cogongrass

September

Ted Bodner

May

Nancy Loewenstein

March

Charles Bryson

Chris Evans

January

Flowering May

Nancy Loewenstein

Seeding May

Nancy Loewenstein

May

Nancy Loewenstein

September

Imperata cylindrica (L.) P. Beauv. **IMCY** **Poaceae**
Synonym: *I. cylindrica* var. *major*
Other common names: japgrass, speargrass, Japanese blood grass ('Red Baron' and other red varieties)

Plant. Aggressive, colony-forming dense perennial grass 1 to 6 feet (30 to 150 cm) in height, often leaning in mats when over 3 feet (90 cm) in height. Stemless tufts of long leaves, blades yellow green, with off-center midveins. Silver-plumed flower and seed heads in late winter (south) through early summer (north). Plants arising from branching sharp-tipped white-scaly rhizomes. Federal noxious weed.

Stem (culm). Upright to ascending, stout, not apparent, as hidden by overlapping leaf sheaths that are long hairy or not.

Leaves. Mainly arising from near the base, long lanceolate, 1 to 6 feet (30 to 180 cm) long and 0.5 to 1 inch (12 to 25 mm) wide, shorter upward. Overlapping sheaths, with outer sheaths often long hairy and hair tufts near the throat. Blades flat or cupped inward, bases narrowing, tips sharp and often drooping. Most often yellowish green. White midvein on upper surface slightly-to-mostly off center (varies in an area). Margins translucent and minutely serrated (rough to touch). Ligule a fringed membrane to 0.04 inch (1.1 mm). Tough to break due to high silica content. Tan colored and persisting after winter dieback.

Flowers. February to June and sporadically (or year-round in Florida). Terminal, silky spikelike panicle, 1 to 8 inches (2.5 to 20 cm) long and 0.2 to 1 inch (0.5 to 2.5 cm) wide, cylindrical and tightly branched on a reddish slender stalk. Spikelets paired, each 0.1 to 0.2 inch (3 to 6 mm) long, obscured by tufts of silky silvery-white hairs to 0.07 inch (1.8 mm).

Seeds. May to June. Tiny oblong brown grain, 0.02 to 0.05 inch (0.5 to 1.3 mm) long, released within dense tufts of silvery hairy husks, often in clusters, for wind dispersal. Seeds matured after V-shaped stigma pair at grain tips shrivel and darken.

Ecology. Grows in full sunlight to partial shade, dry to wet soils, and, thus, can invade a range of stands and sites. Often in circular infestations through rapid growth of branching rhizomes that fill friable soils to a depth of 0.6 to 10 feet (0.1 to 3 m) to exclude most other vegetation. Aggressively invades right-of-ways, new forest plantations, open forests, old fields, and pastures. Absent in areas with frequent tillage, but promoted by burning. Colonizes by rhizomes and spreads by wind-dispersed seeds, seeds and rhizomes contaminating soil and hay, and hitchhiking rides on mowing, logging, and other equipment. Seed fertility highly variable across the region. Highly flammable and a severe fire hazard, burning extremely hot especially in winter.

Resembles Johnsongrass, [*Sorghum halepense* (L.) Pers.], purpletop [*Tridens flavus* (L.) Hitchc.], silver plumegrass [*Saccharum alopecuroides* (L.) Nutt.] and sugarcane plumegrass [*S. giganteum* (Walter) Pers.]—all having a distinct stem and none having an off-center midvein. **Also resembles** longleaf woodoats [*Chasmanthium sessiliflorum* (Poir.) Yates], which lacks off-center midveins and silky flowers, having tufts of spiked flowers and seeds along a slender stalk.

History and use. Introduced from Southeast Asia into FL, southern LA, southern AL, and southern GA in the early to mid-1900s. Initially for soil stabilization. Expectations for improved forage unrealized. Federal noxious weed.

Distribution. Found throughout FL, GA, AL, and MS with scattered infestations in SC, east TX, and LA. The current distribution can be checked at www.cogongrass.org.

Cogongrass **4055**

Stephen F. Enloe

Giant Reed

Arundo donax L. **ARDO4** **Poaceae**

Giant Reed **4008**

Plant. Giant reed grass, cornlike stems, thicket forming in distinct clumps to 20 feet (6 m) in height, with gray-green and hairless stems, long-lanceolate alternate leaves jutting from stems and drooping at the ends, and large plumelike terminal panicles. Seed infertile. Spreading from short tuberous rhizomes that form knotty, spreading mats that penetrate deep into the soil. Dried grass remains standing in winter and spring while low and sheltered plants may remain green.

Stem (culm). Somewhat succulent and fibrous, with round cross section to 1 inch (2.5 cm). Solid jointed every 1 to 8 inches (2.5 to 20 cm) and covered by overlapping leaf sheaths. Gray to yellowish green. Initially white pithed and becoming hollow between joints. Old stems sometimes persistent into the following summer.

Leaves. Alternate, cornlike, long lanceolate with both surfaces hairless, clasping stem with conspicuous whitish base. Eighteen to 30 inches (45 to 76 cm) long and 1 to 4 inches (2.5 to 10 cm) wide near base. Margins and ligule membranous (about 1 mm). Midvein whitish near base becoming inconspicuous towards tip. Veins parallel. Sheaths overlapping, hairless, and semiglossy. White and green variegated forms also escape from cultivation.

Flowers. August to September. Terminal erect dense plumes of whorled, stemmed flowers to 36 inches (1 m) long. Husks hairy, membranous with several veins, and greenish to whitish to purplish.

Seeds. October to March. Dense terminal plume, spindle shaped, densely hairy. Grain infertile.

Ecology. Occurs mainly on upland sites as scattered dense clumps along roadsides and forest margins, migrating from old home plantings by displaced rhizome fragments. Persistent infestations by dense branching tuberous rhizome growth. Probable spread by movement of stem and rhizome parts in soil or by road shoulder grading and by running water. Plants believed to be sterile and not producing viable seeds.

Resembles golden bamboo (*Phyllostachys aurea* Carr. ex A. & C. Rivière), another large grasslike plant that is woody in character. **Closely resembles** common reed [*Phragmites australis* (Cav.) Trin. ex Steud.], which has similar large hairy seed heads, but fanned in a loose plume and not erect, and which occurs mainly near swamps, marshes and other wet habitats.

History and use. Introduced from Western Asia, Northern Africa, and Southern Europe in the early 1800s. Ornamental and reeds for musical instruments (woodwinds).

Distribution. Found throughout the region with scattered dense infestations in every State. Especially frequent along highway and roadside margins.

October

Johnsongrass

Sorghum halepense (L.) Pers. **SORHA** **Poace**
Synonym: *Holcus halepensis* L.

Johnsongrass

Plant. Erect, perennial, warm-season grass, 3 to 8 feet (1 to 2.5 m) tall, stout stemmed and branching from the base having long and wide green leaves with prominent wide, white midveins, from scaly rhizomes and prolific roots to form dense stands. Conical-shaped seed heads with whorled thin branches becoming shorter near the top, each terminated by multiple purplish spikelets. **Caution: Plant toxic to grazing animals if fertilized heavily or drought stricken.**

Stem (culm). Stout, hairless, light green, solid with pronounced whitish swollen nodes. Larger stems branch below midplant.

Leaves. Alternate, long-lanceolate with a tapering tip, 8 to 32 inches (20 to 80 cm) long and 0.4 to 1.2 inches (1 to 3 cm) wide, green sometimes tinged maroon with a prominent wide white midvein and rough serrated margins, hairless except for occasional tufts within the flared throat of the whitish, clasping base. The ligule at the leaf base is a prominent white-fringed membrane, 0.25 inch (2 to 5 mm) long.

Flowers. April to November. Open, spreading panicles, 6 to 20 inches (15 to 50 cm) long, with numerous spaced whorls of projecting fine branches, being shorter in the upper portion. Flattened spikelets in pairs, crowded and overlapping at the end of finer branchlets, 1 spikelet stemless and ovoid, the other stemmed and narrow, 0.15 to 0.24 inch (4 to 6 mm) long. Husks shiny and short hairy, green turning straw yellow to reddish brown when mature, tipped with a thread-like awn or not. Tiny stigmas and stamens project and dangle during flowering.

Seeds. May to March. Grain shiny dark reddish brown turning black, 0.15 to 0.22 inch (4 to 5.5 mm) long, released within straw-colored husks or bare.

Ecology. Occurs as dense colonies in old fields and along field margins and right-of-ways to invade new forest plantations, open forests, and forest openings. Highly competitive with planted and natural tree seedlings, and excludes native plants. Persists and colonizes by rhizomes and spreads by seeds. New plants can produce seed in the first year and seed can remain dormant for many years. Each rhizome segment can sprout. Older plants flattened by running water or vehicles can sprout at each stem node.

Resembles several stout grasses when young, while the seed head shape is more similar to the other sorghum species that are crops and the common native grass, purpletop tridens [*Tridens flavus* (L.) Hitchc.], whose leaves have only a thin whitish midvein, leaf base is often reddish tinged, ligule is a hairy fringe, and seeds are maroon on distinctly drooping panicle branches.

History and use. Introduced in the early 1800s and widely planted as a forage grass and still utilized in some locations. Multiple varieties developed that resulted in cold hardiness and a rapid spread northward.

Distribution. Found throughout the region with scattered dense infestations in every State. Especially frequent along highways, roadsides, and in pastures and hayfields to spread into forest margins and openings.

Steven T. Manning

Nepalese Browntop, Japanese Stiltgrass

Ted Bodner

Chris Oswalt

Ted Bodner

Microstegium vimineum (Trin.) A. Camus **MIVI** **Poaceae**
Other common names: Mary's grass, basketgrass, microstegium

Nepalese Browntop, Japanese Stiltgrass 4080

Plant. Sprawling, annual grass, 0.5 to 3 feet (15 to 90 cm) in height. Flat, short leaf blades with offcenter veins. Stems branching near the base and rooting at nodes to form dense and extensive infestations. Dried whitish-tan grass may remain standing or matted in early winter.

Stem (culm). Ascending to reclining, slender and wiry, up to 4 feet (120 cm) long, with alternate branching. Covered by overlapping sheaths with hairless nodes and internodes. Green to purple to brown. Aerial rootlets descend from lower nodes.

Leaves. Alternate (none basal), projecting out from stem, lanceolate to oblanceolate, 2 to 4 inches (5 to 10 cm) long and 0.07 to 0.6 inch (2 to 15 mm) wide. Blades flat, sparsely hairy on both surfaces and along margins. Midvein whitish and off center. Throat collar hairy. Ligule membranous with a hairy margin.

Flowers. July to October. Terminal, thin and spikelike raceme, to 3 inches (8 cm) long. Unbranched or with 1 to 3 lateral branches on an elongated wiry stem. Other thin racemes of self-pollinating flowers enclosed or slightly extending from lower leaf sheaths and flower/seeding before terminal racemes. Spikelets paired, with the outer stemmed and inner sessile.

Seeds. August to December. Husked grain, seed head thin, grain ellipsoid, 0.1 inch (2.8 to 3 mm) long, with terminal seedstalks partially remaining during early winter.

Ecology. Flourishes on alluvial floodplains and streamsides, mostly colonizing flood-scoured banks, due to water dispersal of seed and flood tolerance. Also common at forest edges, roadsides and trailsides as well as damp fields, swamps, lawns and along ditches. Occurs up to 4,000 feet (1200 m) elevation. Very shade tolerant. Consolidates occupation by prolific seeding, with each plant producing 100 to 1,000 seeds that can remain viable in the soil for 3 years. Spreads on trails and recreational areas by seeds hitchhiking on hikers' and visitors' shoes and clothes.

Resembles nonnative crabgrass (*Digitaria* spp.) and native nimblewill (*Muhlenbergia schreberi* J.F. Gmel.), both having broad short leaves, but distinguished from Nepalese browntop by branching seed heads and stout stems. **Also resembles** whitegrass (*Leersia virginica* Willd.), which is a native perennial with flat, compressed seed heads. **Also resembles** wavyleaf basketgrass [*Oplismenus hirtellus* (L.) P. Beauv.] (nonnative invasive) and basketgrass [*O. hirtellus* (L.) P. Beauv. ssp. *undulatifolius* (Ard.) U. Scholz] (native), which form dense stands of similar appearance in similar habitats, but have wavy leaves and widely branching seed heads.

History and use. Native to temperate and tropical Asia, and first identified near Knoxville, TN, around 1919. Ground cover with little wildlife food value.

Distribution. Found throughout the region with frequent and dense infestations in KY, VA, TN, and NC and spreading south through SC, MS, AL, GA, and into the panhandle of FL.

June

Chris Evans

Tall Fescue

April

May

Ted Bodner

July

May

Ted Bodner

January

Schedonorus phoenix (Scop.) Holub. **SCPH** **Poaceae**
Synonyms: *Festuca arundinacea* Shreb., *F. elatior* L. ssp. *arundinacea* (Schreb.) Hack., *Lolium arundinaceum* (Schreb.) S.J. Darbyshire, *S. arundinaceus* (Schreb.) Dumort
Other common names: meadow fescue, Kentucky 31 fescue

Plant. Erect, tufted, cool-season perennial grass 2 to 4 feet (60 to 120 cm) in height, green in winter and spring during which it is the most common green bunchgrass. Dark-green leaves appearing in late winter, usually flowering in spring (infrequently in late summer). Semidormant during heat of summer, with whitish seedstalks persisting. Growth resuming in fall and continuing into winter. Many cultivars.

Stem (culm). Moderately stout, unbranched, hairless with round cross section and 1 to 3 swollen light-green nodes widely spaced near the base.

Leaves. Mostly basal and a few alternate, flat and long-lanceolate, 4 to 18 inches (10 to 45 cm) long and 0.1 to 0.3 inch (3 to 8 mm) wide. Whitish to yellow-green flared collars, with collar backs often at an angle to the stem. Blades smooth to rough, with 1 to 2 leaves along the stem, becoming smaller upward. Midvein not apparent. Ligule a tiny white membrane.

Flowers. March to June (to October). Loosely branched terminal panicles, 4 to 12 inches (10 to 30 cm) long that are erect or nodding at tips, narrow and tight, then spreading in spring. Branches shorter upward, with 4 to 7 flowers per branch. Flowers greenish white and shiny becoming purplish with whitish stamens and stigma protruding. Spikelets spindle shaped, hairless, ellipsoid with a pointed tip.

Seeds. May (to November). Husked grain, spindle shaped, 0.1 to 0.2 inch (3 to 5 mm) long. Whitish straw-colored husks, usually tipped with a short hair.

Ecology. The predominant cool-season bunchgrass. Occurs as tufted clumps or small to extensive colonies along forest margins and right-of-ways, and widely escaped to invade new forest plantations, roads, openings and high-elevation balds. Grows on wet to dry sites. Spreads by expanding root crowns and less by seeds. Replaces warm-season grassland communities and prairies to the detriment of unique plants and birds. Certain varieties poisonous to livestock and wildlife by infecting them with an endophytic fungus.

Resembles other grasses, especially other fescues and ryegrasses (*Schedonorus* spp.) but distinguished by forming extensive colonies (often planted on roadsides and pastures, to escape into infestations), and having long rounded stems with lower swollen nodes and whitish, flared collars at the base of leaves. Ryegrasses distinguished by producing alternate seed head tight clusters on opposite sides of seedstalks in spring.

History and use. Introduced from Europe in the early to mid-1800s. Recognized as a valuable forage grass in 1930s when the ecotype Kentucky 31 was discovered. Now widely distributed most everywhere in the world. Established widely for turf, forage, soil stabilization and wildlife food plots.

Distribution. Found throughout the region with frequent and dense infestations in KY, VA, TN, NC, MS, and the northern portions of SC and AR.

Tall Fescue **4051**

Ted Bodner

Weeping Lovegrass

John Randall

Eragrostis curvula (Schrad.) Nees **ERCU2** **Poaceae**
Other common name: African lovegrass

Weeping Lovegrass

Plant. Densely clumping to up to 2.5 feet (75 cm) tall, perennial, warm-season grass formed from flattened, basally interconnected sprays of long, thin and wiry basal leaves that arch and droop in all directions almost touching the ground (thus the common name "weeping"). Tall laterally branched flower stalks in early summer to 6 feet (2 m) tall persist during the early winter with seed having varying fertility. Root system is large and fibrous. Evergreen or semi-evergreen in the southern Coastal Plain, while dormant whitish, wispy clumps are highly recognizable further north.

Stem (culm). Not apparent except for the flower/seedstalks. Leaves arise from a tightly packed, short, flattened group of stems hidden in hairy basal sheaths that persist over winter.

Leaves. Thin, less than a quarter of an inch (1 to 4 mm) wide, with margins often rolled inward, to several feet (1 m) long and arching over at midleaf to almost touch the ground. Originating from tightly packed, dense, flattened bundles of leaves, 6 to 12 inches (15 to 30 cm) long encased by persistent sheaths, having scattered to dense hairs.

Flowers. June to July. Open spreading panicles, lavender-gray color, 8 to 10 inches (20 to 25 cm) long, on stalks to 6 feet (1.9 m) tall, with numerous projecting or erect lateral branches, being shorter in the upper portion with numerous secondary branchlets that bear stacks of tiny, grayish-green, husked flowers.

Seeds. July to November. Husks light tan and the grain reddish brown, 0.15 to 0.22 inch (4 to 5.5 mm) long, released within the husks.

Ecology. Still widely seeded for soil stabilization along highways, on surface mines, and around businesses and homes; increasingly escaping to dominate native plant communities throughout the United States. Tolerant of fire, drought, salinity, and bred for cold tolerance. Adapted to a wide range of habitats from moist to dry, hot to cold, and soils that are acid to basic. Prefers well-drained sandy loam soils and will not tolerate standing water. Clumps increase by basal shoots and infestations increase in density by seedfall. Seed dispersed by water, contaminated equipment and soil, and through planting. Occurs as dense colonies in old fields and along field margins and right-of-ways, where it invades new forest plantations, open forests, forest openings and special habitats like native prairies. Detrimental to wildlife, especially ground-nesting and foraging birds.

Resembles no other grasses due to its unique growth habit of tight clumps of "weeping" long and narrow leaves. Many other species of native and nonnative lovegrasses produce similar seedstalks.

History and use. Introduced into the U.S. in 1927 from South Africa and later from Tanzania for erosion control and for forage. Still widely sold and planted as an ornamental and for soil stabilization and forage.

Distribution. Found throughout the region with frequent and dense infestations in KY, VA, TN, and NC and the northern portions of AL, GA, and SC.

February

Japanese Climbing Fern

Ted Bodner

Nancy Loewenstein

Chris Evans

Corrie Pieterson

Lygodium japonicum (Thunb.) Sw. **LYJA** **Lygodiaceae**

Japanese Climbing Fern **5171**

Plant. Perennial viney fern, climbing and twining, to 90 feet (30 m) long, with lacy finely divided fronds along green to orange to black wiry vines or rachis, often forming infestations of shrub- and tree-covering mats. Tan-brown fronds persisting in winter, while others remain green in FL and in sheltered places further north. Vines arising as branches (long compound leaves) from underground, widely creeping rhizomes that are slender, dark brown, and wiry.

Stem (rachis). Slender but difficult to break, twining and climbing, wiry. Green to straw colored or reddish. Mostly deciduous in late winter except in south FL.

Leaves (fronds or pinnae). Opposite on vine, compound, once or twice divided, varying in appearance according to the number of divisions, generally triangular in outline. Three to 6 inches (8 to 15 cm) long and 2 to 3 inches (5 to 8 cm) wide. Highly dissected leaflets, appearing lacy, especially fertile ones. Light green turning tan to dark brown in winter.

Flowers (sporangia). Fertile fronds have smaller segmented fingerlike projections around the margins, bearing sporangia (spore producing dots) in double rows under margins.

Seeds. Late summer to fall (year-round in south FL). Tiny, wind-dispersed spores.

Ecology. Occurs along highway right-of-ways, especially under and around bridges, invading into open forests, forest road edges, and stream and swamp margins. Scattered in open timber stands and plantations, but can increase in cover to form mats, especially after burns, smothering shrubs and trees. Creates "fire ladders" to carry fires upward to scorch and damage canopies of shrubs and trees. Persists and colonizes by rhizomes and spreads rapidly by wind-dispersed spores that are also transported in pine straw mulch and on clothing. Dies back in late winter in the more northern areas, with dead vines providing a trellis for reestablishment.

Resembles Old World climbing fern [*L. microphyllum* (Cav.) R. Br.] and American climbing fern [*L. palmatum* (Bernh.) Sw.], both of which are distinguished by 5 to 7 palmately lobed, fingerlike fronds. American climbing fern—a native occurring in swamps, streambeds, and ravines—does not spread beyond small areas to form extensive infestations. Old World climbing fern, also introduced, is a major invasive pest in mid- to southern FL and projected to migrate northward.

History and use. Native to Asia and tropical Australia and introduced to FL from Japan in the 1930s. An ornamental still being spread by unsuspecting gardeners.

Distribution. Found in dense infestations in southeast TX; south and central LA; central and north FL; and spreading north through AR, MS, AL, and GA. Scattered infestations further north from ornamental plantings in cities and towns.

July

Alligatorweed

Alternanthera philoxeroides (Mart.) Griseb. **ALPH** **Amaranthaceae**
Synonym: *Achyranthes philoxeroides* (Mart.) Standl.
Other common name: alligator weed

Plant. Perennial, evergreen forb, with hollow round stems and opposite leaves at pronounced nodes. When erect, stem tips produce stalked white cloverlike flowers in upper axils during summer, but no fruit or seeds. Trailing or floating stems form entangled mats to 3 feet (90 cm) deep over hundreds of square feet (m) on water and adjoining land. Horizontal jointed stems up to 1 inch (2.5 cm) in diameter and 30 feet (10 m) long readily branch and root at nodes in water to 6.5 feet (2 m) deep or when next to soil.

Stem. Shiny, succulent and round, often reclining, hollow at internodes with a diaphragm at nodes. Pale green with whitish swollen nodes tinged pinkish to purplish to brownish above with age. Nodes topped with a hairy fringe when young, along with lengthwise, minute, hairy internodal grooves, becoming hairless with age.

Leaves. Opposite from swollen nodes, somewhat succulent and shiny, long lanceolate in summer, 0.8 to 2.7 inches (2 to 7 cm) long and 0.4 to 0.8 inch (1 to 2 cm) wide, tapering to the stem with no petiole, being shorter and blunter in winter. Green to blue green above with pale-green midveins and fine hairs. Whitish green and hairless beneath.

Flowers. April to October. Stalked, 0.5 to 3 inches (1.2 to 8 cm) long, in upper leaf axils, small white, rounded, cloverlike cluster of tiny flowers, 0.5 to 0.7 inch (13 to 18 mm) wide, each flower with 5 minute petals (actually sepals) and yellow centers of anthers.

Fruit and seeds. None yet produced in the U.S.

Ecology. Forms mat infestations in shallow water, along shores, and spreading upland from marshes, lakes, rivers, streams, canals, ditches and wet agricultural soils. Grows in both fresh to slightly brackish waters and on sandy to clay soils. Produces deep mats that prevent other plants from germinating in the spring and overtops aquatic and upland plants to damage wetland wildlife habitats. It spreads rapidly by stem fragments moved by water and rooting at nodes.

Resembles other *Alternanthera* species, both nonnative and native, which have similar flowers but none are stalked like alligatorweed. Also resembles the many knotweeds that inhabit wet soils and shorelines that have alternate leaves.

History and use. Native to South America and introduced into the U.S. about 1900. Invading from south to north.

Distribution. Found throughout the region with scattered dense infestations in wetlands of every State.

Alligatorweed

May

Big Blue Lilyturf, Creeping Liriope

Liriope muscari (Decne.) L.H. Bailey **LIMU6** **Liliaceae (Ruscaceae)**
Synonym: *Ophiopogon muscari* Decne.
Creeping Liriope, *L. spicata* (Thunb.) Lour. **LISP10**
Other common names: creeping lilyturf, monkey grass

Big Blue Lilyturf, Creeping Liriope

Plant. The 2 liriope species are similar and even confused in the ornamental trade, with big blue lilyturf being the more common invasive. Both species form dense, evergreen ground cover of crowded tufts of grasslike but thicker leaves, 6 to 18 inches (15 to 45 cm) high and increasing with plant age. Stalked spikelike racemes of small lavender-to-lilac flowers jut upward in early summer to yield stalks of small, green-to-black berrylike fruit in summer through winter. Aggressively spreads by radiating underground stems (rhizomes) that produce a spaced sequence of vertical, white plant initiates, being connected in a line for a time. Young plants have a filmy basal sheath around leaves and a swollen taproot that becomes fibrous and intertwining, eventually filling the upper 6 to 12 inches (15 to 30 cm) of soil. Roots produce scattered small, fleshy, peanut-shaped corms that can sprout as well.

Stem. Grasslike tufts of initially interconnected plants lacking a central stem, except for the flower/fruit stalks.

Leaves. Grasslike but fleshier, radiating from the soil surface in expanding tufted groups, 2.5 to 7 inches (6 to 18 cm) long and 0.1 inch (2 mm) wide on new plants, up to 18 inches (45 cm) long and 0.4 inch (1 cm) wide on mature plants, widest at the middle and tapering to a blunt tip. Multiple lengthwise, parallel veins with a distinctly indented midvein and slightly thickened, very finely serrated margins. Glossy green being somewhat lighter beneath with whitish bases. Variegated varieties appear less aggressive. Leaf tips turn brown and die back in winter. New leaves grow from the base in spring.

Flowers. June to August. Multiple slender flower stalks, 15 to 14 inches (6 to 36 cm) tall, with the terminal 1.5 to 4.5 inches (4 to 12 cm) having spaced clusters of tiny lavender or violet (to white) flowers, with yellow centers, opening at different times.

Fruit and seeds. August to February. Spherical, green, berrylike drupes turn blue ripening to black purple, 0.2 to 0.3 inch (6 to 8 mm) wide. Have grapelike skin and little to no pulp, contain a single spherical dark seed.

Ecology. Grow in full sun to shade and a range of soils, spread is most rapid on moist, highly organic soils. Many cultivars are widely planted as ornamental ground cover and can escape to nearby forests by seeds. Displace forest plants to form ground-layer monocultures. Pollinated by insects. Spread by bird- and animal-dispersed seeds and soil movement with rhizomes and corms.

Resembles mondo grass (*Ophiopogon* spp.), a much smaller-leaved common ornamental. Can resemble tall nut-rush (*Scleria triglomerata* Michx.), a forest nut sedge that has similar clumps of leaves from the soil that are 16 to 24 inches (40 to 60 cm) long, angular flower/seed stalks tipped by brownish leafy bracts that yield tiny, smooth nutlets, white to gray.

History and use. Native to China and Japan. Two of the most common ornamentals planted for ground cover, with both having several varieties.

Distribution. Found in LA, MS, AL, FL, GA, and SC in scattered infestations.

April

Chinese Lespedeza

September

July

February

July

July

Lespedeza cuneata (Dum. Cours.) G. Don **LECU** **Fabaceae**
Synonym: *L. sericea* (Thunb.) Miq., nom. illeg.
Other common name: sericea lespedeza

Chinese Lespedeza **6053**

Plant. Perennial ascending-to-upright leguminous forb 3 to 6 feet (1 to 2 m) in height, with 1-to-many leafy slender stems often branching at midplant, 3-leaflet leaves, and tiny creamy white flowers. Plant arising from a woody root crown. Dormant brown plants remaining upright during most of the winter.

Stems. Often gray green with lines of hairs along the stem.

Leaves. Alternate, crowded and numerous, 3-leaflet leaves. Each leaflet oblong to linear with a hairlike tip, 0.4 to 0.8 inch (1 to 2 cm) long and 0.1 to 0.3 inch (3 to 8 mm) wide. Green above and dense whitish hairy to light gray green beneath. Hairy petioles 0.2 to 0.6 inch (5 to 15 mm) long, absent for upper leaves. Stipules narrowly linear.

Flowers. July to September. Clusters of 1 to 3 pealike flowers crowded in upper leaf axils. Flowers creamy white with purple marks, 0.1 to 0.3 inch (4 to 7 mm) long and shorter than leaves. Hairy 5-lobed calyx shorter than petals.

Fruit and seeds. October to March. Flat ovate to round single-seeded legume pod 0.12 to 0.15 inch (3 to 4 mm) wide. Pods clustered in terminal axils, scattered along the stem and clasped by persistent sepals. Green becoming tan with tiny hairs, with 1 yellow-to-tan seed.

Ecology. Occurs in new and older forest openings, dry upland woodlands to moist savannas, old fields, right-of-ways and cities. Flood tolerant. Forms dense stands by stems sprouting from root crowns, preventing forest regeneration and land access. Cross- and self-pollinates. Spreads slowly from plantings by seeds that have low germination, but remain viable for decades. Nitrogen fixer.

Resembles slender lespedeza [*L. virginica* (L.) Britton], a native, which grows in tufted clumps instead of infestations, has crowded clusters of pink-purple to violet flowers and somewhat larger leaflets 0.6 to 1.2 inches (1.5 to 3 cm) long, and brown stems. **Also resembles** roundhead lespedeza (*L. capitata* Michx.), also native, which has similar leaves and whitish flowers in round-topped clusters.

History and use. Introduced from Japan (contrary to the common name) in 1899, first near Arlington, VA, and soon afterwards in north-central TN. Benefited from government programs that promoted plantings for erosion control and forage. Still planted for quail food plots, soil stabilization, and grazing. Plant improvement breeding programs still underway.

Distribution. Found throughout the region with frequent and dense infestations in AR, north MS and southwest TN, KY, northeast NC, northern SC, and central AL and GA. Not reported in the forests of central and south FL.

Ted Bodner

Coltsfoot

John Cardina

Jan Samanek

Ohio State Weed Lab Archive

Ohio State Weed Lab Archive

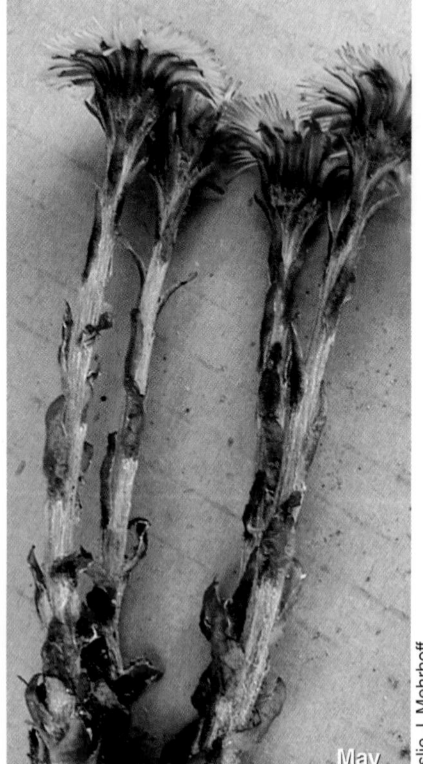

Leslie J. Mehrhoff

Leslie J. Mehrhoff

Chris Evans

Tussilago farfara L. **TUFA** **Asteraceae**

Other common names: horsefoot, foalfoot, assfoot, coughwort, sowfoot

Plant. An unusual low-growing perennial forb from thick branching underground white rhizomes to 10 feet (3 m) deep. Several dandelion-like flower heads per plant sprout in early spring on stout and bracted woolly-haired stalks, then rosettes of colt-hoof shaped leaves appear after dandelion-like plumed seeds have dispersed. The reverse growth sequence of most asters.

Stem. In early spring, several stout stems arise from rhizome tips to 2 to 6 inches (5 to 10 cm) high, covered with woolly-white oppressed hairs and scattered green-to-purple leafy bracts when flowering begins. Stems reach 12 to 20 inches (30 to 50 cm) high by seed dispersal time. Later, leaves on short stems emerge.

Leaves. Arising on short to nonapparent stems near recent seeding stalks, rosettes of small, kidney-shaped leaves give way to long-petioled, flattened, broad heart-shaped leaves, deeply cleft at the base. Both surfaces and petiole initially covered in white, woolly hairs, later to become smooth and glossy on the upper surface, 2 to 7 inches (5 to 18 cm) across and light to dark or bluish green. Many slightly sunken veins radiate from the petiole and extend to the toothed or wavy margin. Purplish-tipped green bracts cling along flower stalks.

Flowers. March to June. Yellow and golden composite heads, 1 to 1.3 inches (2.5 to 3.3 cm) wide, grow on top of bracted stems, each with hundreds of flowers similar to dandelions, enclosed and supported by green-to-purplish tinged sepals (involucre), initially tight and cylindrical, later becoming fully arched back. The outside ring of ray flowers (fertile flowers) extend upward and arch outward turning golden tipped, while inner-disc, center flowers (sterile flowers) are recessed.

Fruit and seeds. June to October. Heads become white fluffy balls, dandelion-like, containing hundreds of tiny thin golden nutlets (achenes) topped with plumes of long white bristles.

Ecology. A severe invasive spreading from the North to the South by wind-dispersed seeds that can travel for miles, although the range may be restricted by summer heat. Can rapidly colonize roadsides, streamsides, and disturbed lands, preferring wet sites but can grow on dry sites, to invade adjacent undisturbed prairies and rocky openings of special habitats to displace threatened species. Rhizomes and seeds can remain dormant in soil for long periods and are stimulated to germinate by disturbance. Seeds germinate throughout the growing season.

Resembles common dandelion (*Taraxacum officinale* L.), which has a smooth flower/seed stalk with white milky sap and long, toothed leaves. Foliage resembles ground ivy (*Glechoma hederacea* L.).

History and use. Introduced from Northern Europe and Asia by early settlers for multiple medicinal uses, although toxic in large doses.

Distribution. Found in scattered infestations throughout TN, KY, NC, and VA.

Coltsfoot

Robert Vidéki

Crownvetch

June

September

Summer

Steven J. Baskauf

June

June

September

Securigera varia (L.) Lassen **SEVA4** **Fabaceae**
Synonym: *Coronilla varia* L.
Other common names: purple crownvetch, trailing crown vetch

Crownvetch

Plant. Deciduous, perennial forb sprawling to form tangled mats to 3 feet (92 cm) high or scrambling over rocks, shrubs, and small trees. Feathery, pinnately compound leaves -similar to true vetch species (*Vicia* spp.) but lacking tendrils- alternate on slender reclining stems. In summer, a multitude of showy, stemmed heads of white and pink flowers jut above entangled plants that yield tufts of slender seedpods in fall. One plant can cover 15 square feet (1.5 m^2) or more in a year from underground stems (rhizomes) with taproots. Forms brown, "earth hugging" patches in winter that resprout quickly in spring or remain green in southern areas. **Caution: All parts are poisonous to some degree.**

Stem. Ascending to sprawling, green, slender and succulent but wiry, slightly angled, 20 to 80 inches (50 to 200 cm) long, with leaves and flower/seed stalks arising along the stem at regular intervals.

Leaves. Alternate, odd-pinnately compound, 2 to 4 inches (5 to 10 cm) long, arising immediately from the stem with 2 tiny stipules. Leaflets 11 to 25, dark green, oblong to obovate, 0.3 to 0.8 inch (0.8 to 2 cm) long with minute hairlike tips.

Flowers. May to September. Small, multicolored pea-type flowers with pink and purple or rose upper petals and white to pinkish-white lower keel petals, 5 to 25 flowers clustered in cloverlike "crowns" about 1 inch (2.5 cm) wide, jutting upward on thin axillary stalks, 2 to 6 inches (5 to 15 cm) long, above entangled plants.

Fruit and seeds. Present May to July and maturing in October. Radiating clusters of slender, pointed seedpods (loments), 2 to 4 inches (5 to 10 cm) long, light green maturing to brown. Segmented to divide and release 3 to 12 flattened brown seeds.

Ecology. Nitrogen fixer and pollinated by insects. Grows in full sun to light shade and the range of conditions common to the South. Tolerates drought, heavy precipitation and cold temperatures. Planted on roadsides, surface mines and in gardens, escaping into forest edges, openings, streamsides and special habitats like rock outcroppings. Displaces plants to form monocultures. Spread by wildlife- and people-dispersed seeds while plant parts buried by sediment along streams can root and grow. Seeds can germinate immediately after release or remain viable in the soil for several years. While reportedly poisonous, especially to horses, it is used for livestock forage. Deer eat crownvetch and it provides cover for rabbits and ground-nesting birds.

Resembles the growth habit and leaves of vetches (*Vicia* spp.) that have tendrils at their leaf ends and scattered flowers, not in clusters like crownvetch.

Distribution. Found in scattered infestations throughout the region.

June

Garlic Mustard

Steven Katovich

John Randall

Chris Evans

Leslie J. Mehrhoff

Garlic Mustard 6002

Alliaria petiolata (M. Bieb.) Cavara & Grande **ALPE4** **Brassicaceae**

Plant. Cool-season biennial, with a slender white taproot, found in small to extensive colonies. Basal rosettes of leaves in the first year remain green during winter and produce 1 to several 2- to 6-foot (60- to 180-cm) tall flower stalks in the second year, and then die after seed formation in midsummer. Dead plants remain standing as long slender seedstalks with many upturned thin seed capsules and a characteristic crook at the stalk base. A faint to strong garlic odor emitted from all parts of the plant when crushed, becoming milder as fall approaches.

Stem. Erect, slightly ridged, light green, hairy lower and hairless above. One to several stems from the same rootstock.

Leaves. Early basal rosette of kidney-shaped leaves and later alternate heart-shaped to triangular leaves, 1.2 to 3.6 inches (3 to 9 cm) long and 1 to 4 inches (2.5 to 10 cm) wide. Margins shallow to coarsely wavy toothed. Tips elongated on stem leaves. Petioles 0.4 to 3 inches (1 to 8 cm) long and reduced upward.

Flowers. April to May (sporadically to July). Terminal, tight clusters of small, white 4-petaled flowers, each 0.2 to 0.3 inch (5 to 7 mm) long and 0.4 to 0.6 inch (10 to 14 mm) wide. Flowering progressing upward as seedpods form below.

Fruit and seeds. May to June. Four-sided, erect-to-ascending, thin pod, 1 to 5 inches (2.5 to 12 cm) long and 0.06 inch (1.5 mm) wide. Initially appearing to be stem branches, spiraled along the stalk. Green, ripening to tan and papery, exploding to expel tiny black seeds arranged in rows.

Ecology. Occurs in small to extensive colonies on floodplains, at forest margins and openings, and less so under dense forest canopies. Shade tolerant while favoring forest edges. Litter disturbance not necessary for establishment. Capable of ballistic seed dispersal of up to 10 feet (3 m). Spreads by human-, animal-, and water-dispersed seeds, which lie dormant for 2 to 6 years before germinating in spring. Experiences year-to-year variations in population densities. Allelopathic, emitting chemicals that kill surrounding plants and microbes.

Resembles violet (*Viola* spp.) in the rosette stage without stalks; and white avens (*Geum canadense* Jacq.) and bittercress (*Cardamine* spp.) that have similar small white flowers, but dissected leaves. None emit garlic odor like garlic mustard.

History and use. Introduced from Europe in the 1800s and first sighted as an escaped weed in 1868 on Long Island, NY. Originally cultivated for medicinal use but no known value now.

Distribution. Found throughout the region except LA, TX, OK, and FL.

Chris Evans

Nodding Plumeless Thistle

Dan Tenaglia

Carduus nutans L **CANU4** **Asteraceae**

Other common names: musk thistle, nodding thistle

Nodding Plumeless Thistle

Plant. A variable biennial or annual herb characterized by sharp spines on leaves, branching stems, and bracts surrounding lavender flowers that sometimes nod to the side (thus the common name). Leafy rosettes first appear in either spring or fall, start forming a deep, hollow taproot (no rhizomes) and bolt within 6 months to 6 feet (2 m) high with midplant branches topped by 1 (to a few) composite flower heads that yield thousands of plumed seeds per plant. The plumes blow away on most, leaving the seed in the head (thus the common name "plumeless"). Spring seedlings can produce seeds in the same year while the more common fall rosettes overwinter to produce flower stalks the next spring.

Stem. Round, fleshy, often covered with dense, white, weblike hairs (or not) towards the tops along with scattered spiny bracts, while spiny leafy ridges extend downward along the more stout and erect branched stems towards the leafy base.

Leaves. Seedling leaves in a rosette are oval to oblong with whitish midveins and margined with tiny spines. Larger rosettes follow with thick, dandelionlike leaves up to 15 inches (40 cm) long, coarsely toothed with whitish, sharp spines extending from each vein. Dark green, hairless, and glossy above with a wide whitish midvein, and lighter green beneath. Similar sized leaves spiral out from the stems and decrease in size upward, becoming leafy spiny bracts scattered below the flower heads.

Flowers. May to September. Each branch topped by a domed flower bud covered with concentric rows of white spine-tipped bracts that part back and arch downward to form a spiny skirt for a showy pinkish to purplish-lavender thistle flower, 1.5 to 3 inches (4 to 8 cm) wide. Each composed of hundreds of tiny perfect flowers, sometimes nodding to the side, and the tallest on a plant being the largest with the most flowers.

Fruit and seeds. June to October. Tightly packed seed heads of tapered nutlets (achenes), each 0.1 to 0.2 inch (3 to 5 mm) long topped by whitish bristles that frequently blow away without the seed. Upwards to 120,000 seeds produced by a single plant in a year.

Ecology. Rosettes have buds that produce sprouts when disturbed. Flowers pollinated by insects, most being cross-pollinated, but self-pollination does occur. Seeds are equipped for dispersal by wind, water, livestock, human activity, and ants, with viability exceeding 10 years in the soil. Most seeds dispersed near the plant and with seed head fall.

Resembles the invasive bull thistle [*Cirsium vulgare* (Savi.) Ten.] that has hairs on the upper leaf surfaces and the perennial Canada thistle [*C. arvense* (L.) Scop.] that has rhizomes, does not appear in the rosette form and only the most outer flower/seed head bracts are spiny.

History and use. Introduced from Southern Europe in the early 1900s as an ornamental.

Distribution. Found throughout the region except FL.

Spotted Knapweed

Steve Dewey

Great Smoky Mtns. NP Res. Mgmt. Archive

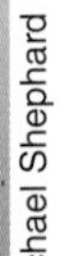

Michael Shephard

John Cardina

Centaurea stoebe L. ssp. **_micranthos_** (Gugler) Hayek **CESTM** **Asteraceae**

Synonym: *C. biebersteinii* DC

Other common name: bushy knapweed

Spotted Knapweed

Plant. A bushy, winter-hardy, upright perennial forb living 3 to 5 years or sometimes longer, often in dense infestations. A deep taproot supports an initial rosette of bluish-green, woolly, dandelion-like leaves. Stem leaves pinnately dissected becoming smaller and less dissected toward the tips of multiple woolly, hairy stems. Midplant branches topped by a few to many pink-to-lavender thistle flowers constricted below the plume by distinctively fringed bracts with black tips (thus the common name "spotted") to produce thousands of tiny bristle-topped seeds. Dead tops remain in winter with new sprouts in spring. A severe invasive in much of the U.S. and now invading the South.

Stem. Round, upright in multistemmed clumps, to 3.2 feet (1 m) tall, covered in dense woolly hairs and having ridges that extend downward from leaf bases. Upper stems wiry and slender with many alternate branches that end in flowers. Young plants may have only 1 stem with 1 flower while older plants can have hundreds of flower-tipped branches.

Leaves. Rosette leaves bluish green, hairy and covered with shiny specks interspersed with translucent dots, 4 to 8 inches (10 to 20 cm) long decreasing in size above midstem, alternate, spiraling and jutting out and upward. Basal leaves deeply divided into elliptic or linear lobes that can appear like leaflets along a wide whitish-to-purplish midvein. Lobes become more slender and fewer on upper leaves, with simple leaves near the flowers.

Flowers. June to November. Each branch topped by an egg-shaped flower bud covered with overlapping rows of dry, fringed (not spiny) green bracts with black tips. Bracts eventually part at top to allow a constricted pinkish-to-lavender thistle plume to radiate out and up, 0.75 to 1 inch (2 to 2.5 cm) wide and overall about 1 inch (2.5 cm) long, composed of hundreds of tiny perfect flowers with the outer ones being sterile.

Fruit and seeds. June to February. Tightly packed seed heads of oblong, brownish, hairy nutlets (achenes), 0.1 inch (3 mm) long, topped by short stubby bristles. Hairs and bristles for clinging. Upwards of 1,000 seeds per plant.

Ecology. Rapidly colonizes roadsides and disturbed lands, especially dry and droughty sites, to invade adjacent undisturbed prairies and open forests. A severe invasive spreading into the South by seeds equipped for dispersal by wind, water, livestock, wildlife, and human activity, with viability in the soil for many years. Seeds germinate throughout the growing season. The roots secrete allelopathic chemicals to inhibit other plant seeds from germinating.

Resembles other thistles and knapweeds, but none have sharp spines, highly dissected leaves with narrow lobes or distinct black spots on the involucre.

History and use. Accidentally introduced from Europe into the Northwestern United States in the 1890s and later spread rapidly across the West and Midwest and into the Northeast and now the South. Listed in most Western States' noxious weed laws.

Distribution. Found throughout the region except OK and TX.

Other Nonnative Plants Invading Southern Forests and Their Margins, Openings, Waterway Margins, Wetlands, and Stream, River, and Lake Banks (species on the Federal Noxious Weed List are denoted by "FED")

Invasive Trees

Common Name:	Scientific Name:
Earleaf acacia (FL only)	*Acacia auriculiformis* A. Cunn. ex Benth.
Amur maple	*Acer ginnala* Maxim.
Norway maple	*Acer platanoides* L.
Woman's tongue (FL and TX only)	*Albizia lebbeck* (L.) Benth.
Deviltree (FL only)	*Alstonia macrophylla* Wall. ex G. Don
Edible fig	*Ficus carica* L.
Chinese parasoltree	*Firmiana simplex* (L.) W. Wight
Paradise apple	*Malus pumila* Mill.
Melaleuca (FL and LA only) FED	*Melaleuca quinquenervia* (Cav.) Blake
White mulberry	*Morus alba* L.
Japanese black pine	*Pinus thunbergii* Parl.
White poplar	*Populus alba* L.
Lombardy poplar	*Populus nigra* L.
Sweet cherry	*Prunus avium* (L.) L.
Cherry plum (TN only)	*Prunus cerasifera* Ehrh.
Sour cherry	*Prunus cerasus* L.
European plum (TX only)	*Prunus domestica* L.
Cherry laurel	*Prunus laurocerasus* L.
Perfumed cherry, Maheleb	*Prunus mahaleb* L.
Sawtooth oak	*Quercus acutissima* Carruthers
Rose myrtle (FL only)	*Rhodomyrtus tomentosus*(Aiton) Hassk.
Octopus tree, schefflera (FL only)	*Schefflera actinophylla* (Endl.) Harms
Peruvian peppertree (TX and FL only)	*Schinus molle* L.
Java plum	*Syzygium cumini* (L.) Skeels
African tamarisk (TX, LA, and SC)	*Tamarix africana* Poir.
Russian tamarisk	*Tamarix aralensis* Willd.
Canary Island tamarisk (LA, GA, SC, and NC)	*Tamarix canariensis* Lour.
Five-stamen tamarisk (TX, OK, AR, and NC)	*Tamarix chinensis* Lour.
French tamarisk, salt-cedar	*Tamarix gallica* L.
Smallflower tamarisk	*Tamarix parviflora* DC.
Saltcedar, tamarisk	*Tamarix ramosissima* Ledeb.
Four-stamen tamarisk (GA only)	*Tamarix tetragyna* C. Ehrenb.
Siberian elm	*Ulmus pumila* L.
Common jujube	*Ziziphus zizyphus* Mill.

Invasive Shrubs

Common Name:	Scientific Name:
Butterfly bush	*Buddleja davidii* Franch.
Surinam cherry	*Eugenia uniflora* L.
Rose of Sharon	*Hibiscus syriacus* L.
Lantana	*Lantana camara* L.
White leadtree	*Leucaena leucocephala* (Lam.) de Wit
Puerto Rico sensitive-briar (TX only)	*Mimosa asperata* L.
Lollipop mimosa, catclaw mimosa (FL and LA only)	*Mimosa pellita* Kunth ex Willd.
Tree tobacco	*Nicotiana glauca* Graham
Guava (FL and LA only)	*Psidium guajava* L.
Scarlet firethorn	*Pyracantha coccinea* M. Roem.
Formosa firethorn	*Pyracantha koidzumii* (Hayata) Rehder
Common pear	*Pyrus communis* L.
Rose myrtle (FL only)	*Rhodomyrtus tomentosus* (Aiton) Hassk.
Jetbead	*Rhodotypos scandens* (Thunb.) Makino
Castorbean	*Ricinus communis* L.
Wineberry, wine raspberry	*Rubus phoenicolasius* Maxim.
Linden arrowwood (VA only)	*Viburnum dilatatum* Thunb.

Invasive Vines

Common Name:	Scientific Name:
Coral vine	*Antigonon leptopus* Hook. & Arn.
Sweet autumn virginsbower	*Clematis terniflora* DC.
Japanese hops	*Humulus japonicus* Siebold & Zucc.
Gloria de la manana	*Ipomoea carnea* Jacq. ssp. *fistulosa* (Mart. ex Choisy) D. Austin
Redstar, red morning-glory	*Ipomoea coccinea* L.
Ivyleaf morning-glory	*Ipomoea hederacea* Jacq.
Tall morning-glory	*Ipomoea purpurea* (L.) Roth
Catclawvine	*Macfadyena unguis-cati* (L.) A.H. Gentry
Stinkvine, skunk vine	*Paederia foetida* L.
Jamaican feverplant (TX, LA, GA, FL)	*Tribulus cistoides* L.
Puncturevine	*Tribulus terrestris* L.

Other Nonnative Plants Invading Southern Forests and Their Margins, Openings, Waterway Margins, Wetlands, and Stream, River, and Lake Banks (species on the Federal Noxious Weed List are denoted by "FED") *(continued)*

Invasive Grasses, Canes, and Grasslikes

Common Name:	Scientific Name:
Jointed goatgrass	*Aegilops cylindrica* Host
Crested wheatgrass	*Agropyron cristatum* (L.) Gaertn.
Desert wheatgrass (TX only)	*Agropyron desertorum* (Fisch. ex Link) J.A. Schult.
Colonial bentgrass	*Agrostis capillaris* L.
Redtop	*Agrostis gigantea* Roth
Tall oatgrass	*Arrhenatherum elatius* (L.) Beauv. ex J. & K. Presl
Basket grass, small carpgrass	*Arthraxon hispidus* (Thunb.) Makino
Animated oat, wild oat	*Avena sterilis* L.
Hedge bamboo (FL only)	*Bambusa multiplex* (Lour.) Raeusch. ex Schult. & Schult. f.
Common bamboo (FL and SC only)	*Bambusa vulgaris* Schrad. ex J.C. Wendl.
Yellow bluestem, King Ranch bluestem	*Bothriochloa ischaemum* (L.) Keng var. *songarica* (Rupr. ex Fisch. & C.A. Mey.) Celarier & Harlan
Field brome, Japanese brome	*Bromus arvensis* L.
Rescuegrass	*Bromus catharticus* Vahl
Smooth brome	*Bromus inermis* Leyss.
Bald brome, meadow brome	*Bromus racemosus* L.
Rye brome	*Bromus secalinus* L.
Cheatgrass	*Bromus tectorum* L.
Japanese sedge	*Carex kobomugi* Ohwi
Bermudagrass	*Cynodon dactylon* (L.) Pers.
Yellow nutsedge	*Cyperus esculentus* L.
Umbrella plant	*Cyperus involucratus* Rottb
Purple nutsedge	*Cyperus rotundus* L.
Orchardgrass	*Dactylis glomerata* L.
Kleberg's bluestem	*Dichanthium annulatum* (Forssk.) Stapf
Angleton bluestem	*Dichanthium aristatum* (Poir.) C.E. Hubbard
African couchgrass	*Digitaria abyssinica* (A. Rich) Stapf
Southern crabgrass	*Digitaria ciliaris* (Retz.) Koel
Hairy or large crabgrass	*Digitaria sanguninalis* (L.) Scop.
Barnyardgrass	*Echinochloa crus-galli* (L.) Beauv.
Indian goosegrass	*Eleusine indica* (L.) Gaertn.
Quackgrass	*Elymus repens* (L.) Gould

Invasive Grasses, Canes, and Grasslikes *(continued)*

Common Name	Scientific Name
Stinkgrass	*Eragrostis cilianensis* (All.) Vign. ex Janchen
Centipede grass	*Eremochloa ophiuroides* (Munro) Hack.
Limpograss	*Hemarthria altissima* (Poir.) Stapf & C.E. Hubbard
Common velvetgrass	*Holcus lanatus* L.
Darnel ryegrass	*Lolium temulentum* L.
Rose Netal grass	*Melinis repens* (Willd.) Zizka
Silkreed (FL and s GA only)	*Neyraudia reynaudiana* (Kunth) Keng ex Hitchc.
Rice	*Oryza sativa* L.
Torpedograss	*Panicum repens* L.
Dallisgrass	*Paspalum dilatatum* Poir.
Bahiagrass	*Paspalum notatum* Flueggé
Kodomillet (AL and GA only) FED	*Paspalum scrobiculatum* L.
Vasey's grass	*Paspalum urvillei* Steud.
Buffelgrass	*Pennisetum ciliare* (L.) Link
African feathergrass (TX only) FED	*Pennisetum macrourum* Trinius.
Crimson fountaingrass	*Pennisetum setaceum* (Forsk.) Chiov.
Timothy	*Phleum pratense* L.
Japanese timber bamboo	*Phyllostachys bambusoides* Siebold & Zucc.
Black bamboo	*Phyllostachys nigra* (Lodd. ex Lindl.) Munro
Common reed (native and introduced)	*Phragmites australis* (Cav.) Trin. ex Steud.
Speargrass	*Poa annua* L.
Canada bluegrass	*Poa compressa* L.
Kentucky bluegrass	*Poa pratensis* L.
Rough bluegrass	*Poa trivialis* L.
Itchgrass FED	*Rottboellia cochinchinensis* (Lour.) W.D. Clayton
Japanese bristlegrass	*Setaria faberi* Herrm.
Foxtail bristlegrass	*Setaria italica* (L.) P. Beauv.
Yellow bristlegrass	*Setaria pumila* (Poir.) Roem. & Schult.
Green bristlegrass	*Setaria viridis* (L.) P. Beauv.
Guineagrass	*Urochloa maxima* (Jacq.) R. Webster
Para grass	*Urochloa mutica* (Forssk.) T.Q. Nguyen
Panic liverseed grass	*Urochloa panicoides* P. Beauv.

Other Nonnative Plants Invading Southern Forests and Their Margins, Openings, Waterway Margins, Wetlands, and Stream, River, and Lake Banks (species on the Federal Noxious Weed List are denoted by "FED") *(continued)*

Invasive Forbs

Common Name:	Scientific Name:
Rosarypea	*Abrus precatorius* L.
Japanese chaff flower	*Achyranthes japonica* (Miq.) Nakai
Russian knapweed	*Acroptilon repens* (L.) DC.
Common bugle	*Ajuga reptans* L.
Wild garlic	*Allium vineale* L. ssp. *vineale*
Sessile joyweed FED	*Alternanthera sessilis* (L.) R. Br. ex DC.
White moneywort	*Alysicarpus vaginalis* (L.) DC.
Thymeleaf sandwort	*Arenaria serpyllifolia* L.
Mugwort, common wormwood	*Artemisia vulgaris* L.
Onionweed FED	*Asphodelus fistulosus* L.
Garden yellowrocket	*Barbarea vulgaris* Ait. f.
Clubed begonia	*Begonia cucullata* Willd.
Corn gromwell	*Buglossoides arvensis* (L.) I.M. Johnston
Basketplant	*Callisia fragrans* (Lindl.) Woods
Sacatrapo,Texasweed	*Caperonia palustris* (L.) A. St.-Hil.
Shepherd's purse	*Capsella bursa-pastoris* (L.) Medik.
Hoary cress, whitetop	*Cardaria draba* (L.) Desv.
Spiny plumeless thistle	*Carduus acanthoides* L.
Garden cornflower	*Centaurea cyanus* L.
White knapweed	*Centaurea diffusa* Lam.
Brown knapweed	*Centaurea jacea* L.
Maltese star-thistle	*Centaurea melitensis* L.
Yellow star-thistle	*Centaurea solstitialis* L.
Alpine knapweed	*Centaurea transalpina* Schleich. ex DC.
Lambsquarters (native and introduced)	*Chenopodium album* L.
Mexican tea	*Chenopodium ambrosioides* L.
Rush skeletonweed	*Chondrilla juncea* L.
Chicory	*Cichorium intybus* L.
Canada thistle	*Cirsium arvense* (L.) Scop.
Bull thistle	*Cirsium vulgare* (Savi) Ten.
Blessed thistle	*Cnicus benedictus* L.
Coco yam, taro	*Colocasia esculenta* (L.) Schott
Benghal dayflower, tropical spiderwort FED	*Commelina benghalensis* (L.) Schott

Invasive Forbs *(continued)*

Common Name	Scientific Name
Asiatic dayflower, common dayflower	*Commelina communis* L.
Poison hemlock	*Conium maculatum* L.
Field bindweed	*Convolvulus arvensis* L.
Piedmont bedstraw	*Cruciata pedemontana* (Bellardi) Ehrend.
Common crupina FED	*Crupina vulgaris* Cass.
Dodder (other than native or widely distrubuted spp.)	*Cuscuta* spp.
Houndstongue, gypsyflower	*Cynoglossum officinale* L.
Jimsonweed	*Datura stramonium* L.
Queen Anne's lace	*Daucus carota* L.
Deptford pink	*Dianthus armeria* L.
Fuller's teasel	*Dipsacus fullonum* L.
Common teasel	*Dipsacus fullonum* ssp. *sylvestris* (Huds.) Clapham
Cut-leaf teasel	*Dipsacus laciniatus* L.
Indian strawberry	*Duchesnea indica* (Andr.) Focke
Redstem stork's bill	*Erodium cicutarium* (L.) L'Hér. ex Ait.
Leafy spurge	*Euphorbia esula* L.
Mullberrryweed, hairy crabweed	*Fatoua villosa* (Thunb.) Nakai
Sweet fennel	*Foeniculum vulgare* P. Mill.
Goatsrue	*Galega officinalis* L.
South American mock vervain	*Glandularia pulchella* (Sweet) Troncoso
Fulva, common orange daylily	*Hemerocallis fulva* (L.) L.
Giant hogweed FED	*Heracleum mantegazzianum* Sommier & Levier
Dames rocket	*Hesperis matronalis* L.
Orange hawkweed	*Hieracium aurantiacum* L.
Meadow hawkweed	*Hieracium caespitosum* Dumort.
Common St. Johnswort	*Hypericum perforatum* L.
Paleyellow iris	*Iris pseudacorus* L.
Korean clover	*Kummerowia stipulacea* (Maxim.) Makino
Japanese clover	*Kummerowia striata* (Thunb.) Schindl.
Willowleaf lettuce	*Lactuca saligna* L.
Prickly lettuce	*Lactuca serriola* L.

continued

Other Nonnative Plants Invading Southern Forests and Their Margins, Openings, Waterway Margins, Wetlands, and Stream, River, and Lake Banks (species on the Federal Noxious Weed List are denoted by "FED") *(continued)*

Invasive Forbs *(continued)*

Common Name:	Scientific Name:
Henbit deadnettle	*Lamium amplexicaule* L.
Common nipplewort	*Lapsana communis* L.
Common motherwort	*Leonurus cardiaca* L.
Oxeye daisy	*Leucanthemum vulgare* Lam.
Yellow toadflax	*Linaria vulgaris* P. Mill.
Bird's-foot trefoil	*Lotus corniculatus* L.
Horehound	*Marrubium vulgare* L.
Black medick	*Medicago lupulina* L.
Burclover	*Medicago polymorpha* L.
White sweet clover	*Melilotus alba* Ders.
Yellow sweet clover	*Melilotus officinalis* (L.) Lam.
Melilotus, clover	*Melilotus* spp.
Miniature beefsteakplant	*Mosla dianthera* (Buch.-Ham. ex Roxb.) Maxim.
Catnip	*Nepeta cataria* L.
Scotch thistle	*Onopordum acanthium* L.
Sleepydick, Star-of-Bethlehem	*Ornithogalum umbellatum* L.
Hellroot, small broomrape FED	*Orobanche minor* Smith
Wild parsnip	*Pastinaca sativa* L.
Harmal peganum	*Peganum harmala* L.
Beefsteakplant	*Perilla frutescens* (L.) Britt.
Mascarene Island leaf-flower	*Phyllanthus tenellus* Roxb.
Chamber bitter	*Phyllanthus urinaria* L.
Narrowleaf plantain	*Plantago lanceolata* L.
Asiatic smartweed, Oriental ladysthumb	*Polygonum caespitosum* Blume
Asiatic tearthumb	*Polygonum perfoliatum* L.
Spotted ladysthumb	*Polygonum persicaria* L.
Giant knotweed	*Polygonum sachalinense* F. Schmidt ex Maxim.
Sulfur cinquefoil	*Potentilla recta* L.
St. Anthony's turnip	*Ranunculus bulbosus* L.
Fig buttercup, lesser celandine	*Ranunculus ficaria* L.
Wild radish	*Raphanus raphanistrum* L.
Cultivated radish	*Raphanus sativus* L.
Annual bastardcabbage	*Rapistrum rugosum* (L.) All.
Creeping yellowcress	*Rorippa sylvestris* (L.) Bess.

Invasive Forbs *(continued)*

Common Name	Scientific Name
Britton's wild petunia	*Ruellia caerulea* Leonard
Common sheep sorrel	*Rumex acetosella* L.
Curly dock	*Rumex crispus* L.
Prickly Russian thistle	*Salsola tragus* L.
Shrubby Russian thistle	*Salsola vermiculata* L.
South American skullcap	*Scutellaria racemosa* Pers.
Sicklepod	*Senna obtusifolia* (L.) Irwin & Barneby
Rattlebush, rattlebox	*Sesbania punicea* (Cav.) Benth.
London rocket	*Sisymbrium irio* L.
Twoleaf nightshade	*Solanum diphyllum* L.
Turkeyberry	*Solanum torvum* Sw.
Perennial sowthistle, field sowthistle	*Sonchus arvensis* L.
Spiny sowthistle	*Sonchus asper* (L.) Hill
Common sowthistle	*Sonchus oleraceus* L.
Slender sowthistle	*Sonchus tenerrimus* L.
Bay Biscayne creeping-oxeye, wedelia	*Sphagneticola trilobata* (L.C. Rich.) Pruski
Florida hedgenettle	*Stachys floridana* Shuttlw. ex Benth.
Common chickweed	*Stellaria media* (L.) Vill. ssp. *media*
Common chickweed	*Stellaria media* (L.) Vill. ssp. *pallida* (Dumort.) Asch. & Graebn.
Witchweed	*Striga asiatica* (L.) Kuntze
Cowpea witchweed	*Striga gesnerioides* (Willd.) Vatke
Common dandelion	*Taraxacum officinale* G.H. Weber ex Wiggers
Spreading hedgeparsley	*Torilis arvensis* (Huds.) Link
Yellow salsify	*Tragopogon dubius* Scop.
Coatbuttons	*Tridax procumbens* L.
White clover	*Trifolium repens* L.
Clover	*Trifolium* spp.
Stinging nettle (native and introduced)	*Urtica dioica* L.
Common mullein	*Verbascum thapsus* L.
Purpletop vervain	*Verbena bonariensis* L.
Brazilian vervain	*Verbena brasiliensis* Vell.
Ivyleaf speedwell	*Veronica hederifolia* L.
Garden vetch	*Vicia sativa* L.
Spiny cocklebur	*Xanthium spinosum* L.

Other Nonnative Plants Invading Southern Forests and Their Margins, Openings, Waterway Margins, Wetlands, and Stream, River, and Lake Banks (species on the Federal Noxious Weed List are denoted by "FED") *(continued)*

Invasive Ferns

Common Name:	Scientific Name:
Asian netvein hollyfern	*Cyrtomium fortunei* J. Sm.
Japanese false spleenwort	*Deparia petersenii* (Kunze) M. Kato
Old world climbing fern	*Lygodium microphyllum* (Cav.) R. Br.
Narrow swordfern, Boston fern	*Nephrolepis cordifolia* (L.) C. Presl
Chinese brake fern, ladder brake	*Pteris vittata* L.

Aquatic and Wetland Plants, Ferns, and Algae

Common Name:	Scientific Name:
Feathered mosquitofern (only in NC) FED	*Azolla pinnata* R. Brown
Brazilian water-hyssop	*Bacopa egensis* (Poepp.) Pennell
Pond water-starwort	*Callitriche stagnalis* Scop.
Water sprite	*Ceratopteris thalictroides* (L.) Brongn.
Becket's water trumpet	*Cryptocoryne beckettii* Thwaites ex Trimen.
Rock snot, didymo	*Didymosphenia geminata* (alga) (Lyngbye) M. Schmidt
Brazilian egeria	*Egeria densa* Planch.
Common water hyacinth	*Eichhornia crassipes* (Mart.) Solms
Waterthyme, hydrilla FED	*Hydrilla verticillata* (L. f.) Royle
Water-poppy	*Hydrocleys nymphoides* (Humb. & Bonpl. ex Willd.) Buchenau
Indian swampweed, miramar weed FED	*Hygrophila polysperma* (Roxb.) T. Anders.
Yellow flag iris	*Iris pesudacorus* L.
Dotted duckweed, dotted duckmeat	*Landoltia punctata* (G. Mey.) D.H. Les & D.J. Crawford
Asian marshweed (FL, GA, and TX) FED	*Limnophila sessiliflora* (Vahl) Blume
Large-flower primrose-willow	*Ludwigia grandiflora* (Michx.) Greuter & Burdet spp. *grandiflora*
Creeping or six-petal water primrose	*Ludwigia grandiflora* (Michx.) Greuter & Burdet spp. *hexapetala* (Hook. & Arn.) G.L. Nesom & Kartesz
Moneywort, creeping Jenny	*Lysimachia nummularia* L.

Aquatic and Wetland Plants, Ferns, and Algae *(continued)*

Common Name	Scientific Name
Purple loosestrife	*Lythrum salicaria* L.
European wand loosestrife (VA only)	*Lythrum virgatum* L.
Dwarf waterclover	*Marsilea minuta* L.
Australian waterclover	*Marsilea mutica* Mett.
European waterclover	*Marsilea quadrifolia* L.
Peppermint	*Mentha* ×*piperita* L. (pro sp.) [aquatica × spicata]
Asian spiderwort, marsh dewflower	*Murdannia keisak* (Hassk.) Hand.-Maz.
Parrots feather watermilfoil	*Myriophyllum aquacticum* (Vell.) Verdc.
Eurasian watermilfoil	*Myriophyllum spicatum* L.
Brittle waternymph	*Najas minor* All.
Watercress	*Nasturtium officinale* Ait. f.
Sacred lotus	*Nelumbo nucifera* Gaertn.
Cape Blue water-lily	*Nymphaea capensis* Thunb.
Yellow floatingheart	*Nymphoides peltata* (Gmel.) Kuntze
Duck-lettuce FED	*Ottelia alismoides* (Linnaeus) Pers.
Cuban bulrush	*Oxycaryon cubense* (Poepp. & Kunth) Lye
Common reed	*Phragmites australis* (Cav.) Trin. ex Steud.
Water lettuce	*Pistia stratiotes* L.
Curly pondweed	*Potamogeton crispus* L.
Giant arrowhead	*Sagittaria montevidensis* Cham. & Schltdl.
Waterspangles, common salvinia	*Salvinia minima* Baker
Kariba-weed, giant salvinia	*Salvinia molesta* Mitchell
Salvinia	*Salvinia* spp.
Bog bullrush	*Schoenoplectus mucronatus* (L.) Palla
Bigpod sesbania	*Sesbania herbacea* (Mill.) McVaugh
Wetlands nightshade	*Solanum tampicense* Dunal
Exotic bur-reed FED	*Sparganium erectum* L.
Water chestnut	*Trapa natans* L.
European speedwell	*Veronica beccabunga* L.

Sources of Identification Information

Books

Dirr, M.A. 1975. Manual of woody landscape plants. Revised. Champaign, IL: Stripes Publishing. 1187 p.

Godfrey, R.K. 1988. Trees, shrubs, and woody vines of northern Florida and adjacent Georgia and Alabama. Athens, GA: The University of Georgia Press. 734 p.

Kaufman, S.R.; Kaufman, W. 2007. Invasive plants; guide to identification and the impacts and control of common North American species. Mechanicsburg, PA: Stackpole Books. 458 p.

Langeland, K.A.; Burks, K.C., ed. 1998. Identification & biology of non-native plants in Florida's natural areas. Gainesville, FL: University of Florida. 165 p.

Miller, J.H.; Miller, K.V. 2005. Forest plants of the Southeast and their wildlife uses. Athens, GA: The University of Georgia Press. 454 p.

Randall, J.M.; Marinelli, J., ed. 1996. Invasive plants: weeds of the global garden. Handb. 149. Brooklyn, NY: Brooklyn Botanic Garden. 111 p.

Weakley, A.S. 2006. Flora of the Carolinas, Virginia, Georgia, and surrounding areas (working draft of 17 January 2006). Chapel Hill, NC: University of North Carolina Herbarium (NCU), North Carolina Botanical Garden, and the University of North Carolina at Chapel Hill. 1026 p.

Manuals

Smith, Tim E., ed. 1993. Missouri vegetation management manual. Jefferson City, MO: Missouri Department of Conservation, Natural History Division. 148 p.

Swearingen, J.; Reshetiloff, K.; Slattery, B.; Zwicker, S. 2002. Plant invaders of mid-Atlantic natural areas. Washington, DC: National Park Service; U.S. Fish & Wildlife Service. 82 p.

Tennessee Exotic Pest Plant Council. 1996. Tennessee exotic plant management manual. Nashville, TN: Tennessee Exotic Pest Plant Council, Warner Parks Nature Center. 118 p.

Articles and Reports

Bruce, K.A.; Cameron, G.N.; Harcombe, P.A.; Jubinsky, G. 1997. Introduction, impact on native habitats, and management of a woody invader, the Chinese tallow-tree, *Sapium sebiferum* (L.) Roxb. Natural Areas Journal. 17: 255–260.

Culley, T.M.; Hardiman, N.A. 2007. The beginning of a new invasive plant: a history of the ornamental callery pear in the United States. BioScience. 57: 956–964.

Forseth, I.N.; Innis, A.F. 2004. Kudzu (*Pueraria montana*): history, physiology, ecology combine to make a major ecosystem threat. Critical Reviews in Plant Sciences. 23: 401–413.

Mueller, T.C.; Robinson, D.K.; Beller, J.E. [and others]. 2003. *Dioscorea oppositifolia* L. phenotypic evaluation and comparison of control strategies. Weed Technology. 17: 705–710.

Mullahey, J.J.; Colvin, D.L. 2000. Weeds in the sunshine: tropical soda apple (*Solanum viarum*) in Florida—1999. Gainesville, FL: University of Florida, Institute of Food and Agricultural Sciences. 7 p.

Schierenbeck, K.A. 2004. Japanese honeysuckle (*Lonicera japonica*) as an invasive species; history, ecology, and context. Critical Reviews in Plant Science. 23: 391–400.

Trusty, J.L.; Goertzen, L.R.; Zipperer, W.C.; Lockaby, B.G. 2007. Invasive wisteria in the Southeastern United States: genetic diversity, hybridization and role of urban centers. Urban Ecosystems. 10: 379–395.

Trusty, J.L.; Lockaby, B.G.; Zipperer, W.C.; Goertzen, L.R. 2007. Identity of naturalized exotic wisteria (Fabaceae) in the South-eastern United States. Weed Research. 47: 479–487.

Vincent, M.A. 2005. On the spread and current distribution of *Pyrus calleryana* in the United States. Castanea. 70: 20–31.

Newsletters and Magazines

Florida Exotic Plant Pest Council. 1996–2008. Wildland weeds. Gainesville, FL. Quarterly.

Web Sites

Global

Global Invasive Species Database: http://www.issg.org/database/welcome/

An International Nonindigenous Species Database Network: http://www.nisbase.org/nisbase/index.jsp

The Nature Conservancy's Global Invasive Species Team: http://www.invasive.org/gist/

National

Center for Invasive Species and Ecosystem Health: http://www.bugwood.org/index.cfm

Institute for Biological Invasion: http://invasions.bio.utk.edu/Default.htm

Invasive and Exotic Species: http://www.invasive.org/

National Association of EPPCs: http://www.naeppc.org/

NatureServe: http://www.natureserve.org/explorer/

Federal Government Agencies

Alien Plant Invaders of Natural Areas (PCA, National Park Service): http://www.nps.gov/plants/alien/factmain.htm

APHIS: http://www.aphis.usda.gov/plant_health/plant_pest_info/weeds/index.shtml

ARS: Exotic & Invasive Weed Research: http://www.ars.usda.gov/main/site_main.htm?modecode=53-25-43-00

CSREES Invasive Species (USDA Cooperative Extension Service): http://www.csrees.usda.gov/invasivespecies.cfm

A Field Guide for Identification and Control, PLANTS National Database: http://plants.usda.gov/

National Institute of Invasive Species Science: http://www.niiss.org/cwis438/websites/niiss/About.php?WebSiteID=1

NBII Invasive Species Information Node: http://invasivespecies.nbii.gov/

Nonnative Invasive Plants of Southern Forests: http://www.srs.fs.usda.gov/pubs/gtr/gtr_srs062/index.htm

Nuisance Species Task Force (ANS): http://www.anstaskforce.gov/default.php

Southern Region, Forest Service, Forest Inventory and Analysis Invasive Plants: http://srsfia2.fs.fed.us/nonnative_invasive/southern_nnis.php

The U.S. National Arboretum: www.usna.usda.gov/Gardens/invasives.html

U.S. Army Environmental Center (USAEC) Pest Management: http://el.erdc.usace.army.mil/pmis/

USDA ARS Invaders Database System: http://invader.dbs.umt.edu/Noxious_Weeds/

USDA Forest Service: http://www.fs.fed.us/foresthealth/programs/invasive_species_mgmt.shtml

USDA Forest Service Invasive Species Program: http://www.fs.fed.us/invasivespecies/index.shtml

USDA National Invasive Species Information Center: http://www.invasivespeciesinfo.gov/

U.S. Environmental Protection Agency Invasive Species Program: http://www.epa.gov/owow/invasive_species/

U.S. Fish and Wildlife Service Invasive Species Program: http://www.fws.gov/invasives/

USGS Nonindigenous Aquatic Species: http://nas.er.usgs.gov/

Weeds Gone Wild: Alien Plant Invaders of Natural Areas: http://www.nps.gov/plants/alien/

Regional

Cogongrass Web site, The University of Georgia: http://www.cogongrass.org

Flora of the Southeast: http://www.herbarium.unc.edu/seflora/species.htm

Invasive Plant Atlas of New England: http://nbii-nin.ciesin.columbia.edu/ipane/index.htm

Mid-Atlantic Exotic Pest Plant Council: http://www.ma-eppc.org/

Mid-West Invasive Plant Network: http://mipn.org/

Southeast EPPC: http://www.se-eppc.org/

State

NASDA (National Association of State Departments of Agriculture): http://www.nasda.org/

State Exotic Pest Plant Councils: http://www.naeppc.org/

State Laws and Regulations: http://www.invasivespeciesinfo.gov/laws/statelaws.shtml

Texasinvasive.org: http://www.texasinvasives.org/

Glossary of Important Terms

achene: a small, dry, nonsplitting fruit with a single seed, common to grasses, asters, and nut-rushes.

acute tip: terminating in a sharp or well-defined point, with more or less straight sides.

allelopathic: referring to a plant known to emit chemicals that retard the growth or seed germination of associated plants.

alternate leaves: one leaf at each node and alternating on sides of the stem or their points of attachment forming a spiral up the stem.

annual: a plant that germinates, flowers, produces seed, and dies within one growing season.

anthers: the pollen-producing portion of the stamen or male reproductive part of a flower.

appressed: lying close to or flattened against.

arbor: vine entanglement within the crowns of shrubs or trees.

ascending: tending to grow upward, slightly leaning to somewhat erect.

asymmetric: not identical on both sides of a central line.

axil: the angle formed between two structures, such as between a leaf and the stem.

axillary: located in an axil or angle.

berry: a fleshy or pulpy fruit from a single ovary with one to many embedded seeds, such as tomato and grape.

biennial: a plant that lives for about 2 years, typically forming a basal rosette in the first year, flowering and fruiting in the second year, and then dying.

bipinnately compound: twice pinnately compound; a pinnately compound leaf being again divided.

blade: the expanded part of a leaf.

bract: a small leaf or leaflike structure at the base of a flower, inflorescence, or fruit.

branch scar: a characteristic marking on a stem where there was once a branch.

bud: an undeveloped flower, flower cluster, stem, or branch, often enclosed by reduced or specialized leaves termed bud-scales.

bulbil: an aerial tuber.

bunch grass: a grass species with a cluster-forming growth habit; a grass growing in an upright large tuft.

bundle scar: tiny raised area(s) within a leaf scar, from the broken ends of the vascular bundles, found along a twig.

calyx: the collective term for all of the sepals of a flower, commonly green, but occasionally colored and petallike or reduced to absent.

calyx tube: a tubelike structure formed by wholly or partially fused sepals.

cane: very tall grasses, for example, switchcane and bamboo; tall, stiff stem.

capsule: a dry fruit that splits into two or more parts at maturity, for example, the fruit of tallowtree.

clasping: base that partly or wholly surrounds another structure, such as a leaf base surrounding a stem.

collar: the area of a grass leaf blade where it attaches to the sheath.

colony: a stand or group of one species of plant, from seed origin or those connected by underground structures such as rhizomes.

cordate: heart shaped.

cordate base: a leaf base resembling the double-curved top of a heart shape.

corolla: the collective name of all of the petals of a flower.

cotyledon: the initial leaves on a plant germinant.

crenate: margin with shallow, rounded teeth; scalloped.

cultivar: a form or variety of plant originating under cultivation.

deciduous: falling off or shedding; not persistent; refers to leaves, bracts, stipules, and stipels.

dioecous: plants with unisexual flowers and having male and female flowers on separate plants.

drupe: a fleshy fruit, surrounding a stone (endocarp) that contains a single seed.

ellipsoid: a three-dimensional ellipse; narrow or narrowly rounded at ends and widest in the middle.

elliptic: oval-shaped; broadest at the middle and rounded and narrower at the two equal ends.

entire: margins without teeth, notches, or lobes.

even pinnately compound: a leaf with two or more leaflets arranged opposite along a leafstalk or rachis.

evergreen: green leaves remaining present through winter.

exotic: foreign; originating on a continent other than North America.

fern: a broadleaf pteridophyte of the order Filicales, typically with much-divided leaves and spore reproduction.

filament: the long, slender stalk of a stamen that supports the anther.

forb: a broad-leaved herbaceous (nonwoody) plant.

frond: a large, once- or twice-divided leaf, here referring to fern leaves.

gland: a structure which contains or secretes a sticky, shiny, or oily substance.

grain: a grass seed.

grass: plants of the family Poaceae, typically with narrow leaves and jointed stems.

hairy: surface features of plants, many protruding filaments or glands that give texture; pubescent.

herb or herbaceous: a plant with no persistent aboveground woody stem, dying back to ground level at the end of the growing season.

hip fruit: the fruit of the genus *Rosa* that is ovoid, fleshy, and usually red when ripe.

husk: the outer scalelike coverings of a grass seed.

inflorescence: the flowering portion of a plant; the flower cluster; the arrangement of flowers on the stem.

internode: the space on an herb or grass stem between points of leaf attachment.

involucre: a whorl or collection of bracts subtending or enclosing a flower or flower cluster.

lanceolate: lance shaped; widest at or near the base and tapering to the apex.

lateral: on or at the sides, as opposed to terminal or basal.

leaflet: an individual or single division of a compound leaf.

leaf scar: the scar or marking left on a twig after leaf fall.

leafstalk: the main stem of a compound leaf, rachis.

legume: a plant in the family Fabaceae; a dry, splitting fruit, one-to-many seeded, derived from a single carpel and usually opening along two sutures, confined to the Fabaceae.

legume pod: the fruit of a legume.

lenticel: a raised dot or short line, usually corky to white in color, on twigs and stems.

ligule: a tiny membranous projection, often fringed with hairs, from the summit of the sheath (top of the throat), where the leaf attaches, in many grasses and some sedges.

linear: long and narrow shaped with roughly parallel sides.

lobed leaf: margins having deep indentations resulting in rounded-to pointed portions.

loments: linear seedpods of the legume family.

margin: the edge of a leaf blade or flower petal; the edge of a forest.

marsh: a poorly drained portion of the landscape with shallow standing water most of the year, most extensive around intertidal zones.

membranous: thin, filmy, and semitransparent.

midvein: the central vein of a leaf or leaflet.

milky sap: sap being opaque white and often of a thick consistency.

monocot: the class of plants having one cotyledon (or monocotyledonous) and parallel leaf veins, including grasses, sedges, lilies, and orchids.

mottled: spotted or blotched in color.

node: the point of leaf or stem attachment, sometimes swollen on grass stems where the sheath is attached.

nutlet: a small, dry, nonsplitting fruit with a woody cover, usually containing a single seed.

oblanceolate: lance shaped with the widest portion terminal; inversely lanceolate.

oblong: a shape two-to-four times longer than wide with nearly parallel sides.

obovate: two-dimensional egg-shaped, with the attachment at the narrow end; inverted ovate.

odd-pinnately compound: pinnately compound leaves with a terminal leaflet rather than a terminal pair of leaflets or a terminal tendril.

opposite: leaves born in pairs at each node on opposite sides of the stem.

ornamental: a plant cultivated for aesthetic purposes.

oval: broadly elliptic in shape, with the width greater than half of the length.

ovate: two-dimensional egg-shaped, with the attachment at the wider end.

ovoid: three-dimensional egg-shaped, with the attachment at the wider end.

palmate: like a hand with fingerlike leaflets or veins radiating outward.

panicle: an irregularly branched inflorescence with the flowers maturing from the bottom upward.

pealike flower: irregular flower characteristic of sweet peas and beans in the family Fabaceae.

perennial: any plant that persists for three or more growing seasons, even though it may die back to rhizomes or rootstock during the dormant period.

petiole: a stalk that attaches the leaf blade to the stem.

pinnately compound: a compound leaf with leaflets arising at intervals along each side of an axis or rachis (leafstalk).

pistil: the female reproductive portion of a flower, usually consisting of an ovary, style, and stigma.

pith: the soft or spongy central tissue in some twigs and stems, sometimes absent making the stem hollow.

plume: a tuft of simple or branched bristles.

pod: an elongated dry fruit that usually splits open upon maturity, such as a legume.

prickle: a spinelike structure protruding from a stem.

raceme: an elongated, unbranched inflorescence with stalked flowers generally maturing from the bottom upward.

rachis: the main axis of an inflorescence or compound leaf.

recurved: gradually curved backward or downward.

rhizome: an underground stem, usually horizontal and rooting at nodes.

right-of-way: a narrow corridor of land in straight sections across the landscape, repeatedly cleared and kept in low vegetation, to accommodate roadway structures, poles and wire for electrical and telephone transmissions, and pipelines.

riparian: situated or dwelling on the bank or floodplain of a river, stream, or other body of water.

root collar: the surface area of a perennial where the stem and roots join.

rootcrown: the part of a perennial plant where the stem and roots join, often swollen.

root sprout: a plant originating from a root or rhizome that takes root at nodes.

rootstock: the part of a perennial plant near the soil surface where roots and shoots originate.

rosette (basal rosette): a circular cluster of leaves on or near the soil surface radiating from a rootcrown, as in dandelions.

scaly: covered with minute flattened, platelike structures.

semi-evergreen: tardily deciduous or maintaining green foliage during winter only in sheltered locations.

semiwoody plants: species that have mostly woody stems and deciduous leaves, usually shorter than shrubs.

sepal: a single unit of the calyx; the lowermost whorl of flower parts.

serrate: margin with sharp forward-pointing teeth.

sessile: attached without a stalk, such as a leaf attached without a petiole.

shade intolerant: a plant that cannot grow and reproduce under the canopy of other plants but needs direct sunlight.

shade tolerant: a plant that can grow and reproduce under the canopy of other plants.

sheath: a more or less tubular portion of a structure surrounding another structure, such as the tubular portion of leaf bases of grasses that surround the stem.

shrub: a wood plant, typically multistemmed and shorter than a tree.

simple: not compound; single; undivided; unbranched.

sinus: a rounded indention in leaves, often liken to mitten shapes.

smooth: not rough to the touch, usually hairless (or only finely hairy) and scaleless.

spherical: round in three dimensions, like a ball; synonymous with globose.

spike: an elongated, unbranched inflorescence with sessile or unstalked flowers along its length, the flowers generally maturing from the bottom upward.

sporangia: the case-bearing spores on ferns.

spore: a minute (almost not visible), one-celled reproductive body of ferns, asexual.

stamen: the male reproductive portion of a flower, usually consisting of an anther and filament.

stigma: tip of the female part of a flower where pollen lands and germinates to fertilize the ovary for seed production.

stipular: originating from stipules.

stipules: the pair of leaflike structures at the base of a leaf petiole in some species.

stone: a hard woody structure enclosing the seed of a drupe.

subshrub: a very short woody plant.

subtend: a structure just below another, such as flowers subtended by bracts.

succulent: fleshy or soft tissued.

swamp: a wooded or brushy area usually having surface water.

synonym: a discarded scientific name for a plant; another common name.

taproot: the main root axis; a long vertical, central root.

tardily deciduous: maintaining at least some green leaves into winter or early spring.

terminal: at the end.

thorn: a stiff, curved, sharply pointed modified stem, sometimes branched.

throat: the area inside a flower tube formed from fused petals; the upper side of a grass collar where the blade meets the sheath.

toothed: margin with outward pointed lobes; coarsely dentate.

trailing: running along the soil or leaf litter surface.

tuber: a thickened portion of a root or rhizome modified for food storage and vegetative propagation, such as a sweet potato.

tubular: a cylindrical structure, such as formed from fused petals or sepals.

twig: short leaf branch.

umbel: a compound flower with stems arising and radiating from one point of attachment.

variegated: marked with stripes or patches of different colors.

vine: a long trailing or climbing plant.

whorled: three or more leaves in a circular arrangement arising from a single node or radiating at different angles to the main stem.

wiry: thin, flexible, and tough.

yam: a tuber or potatolike organ.

FLOWER PARTS

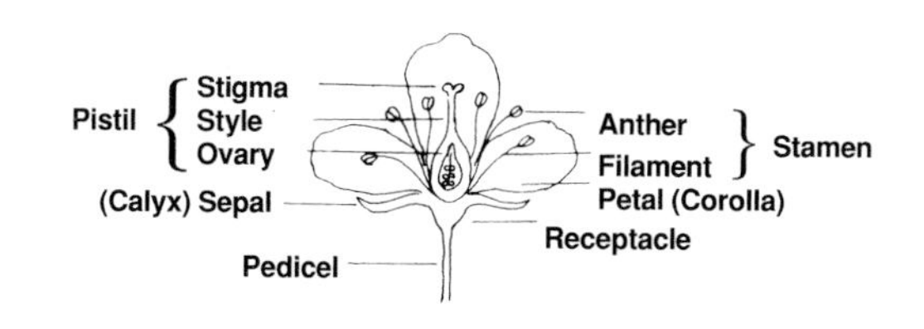

FLOWER TYPES

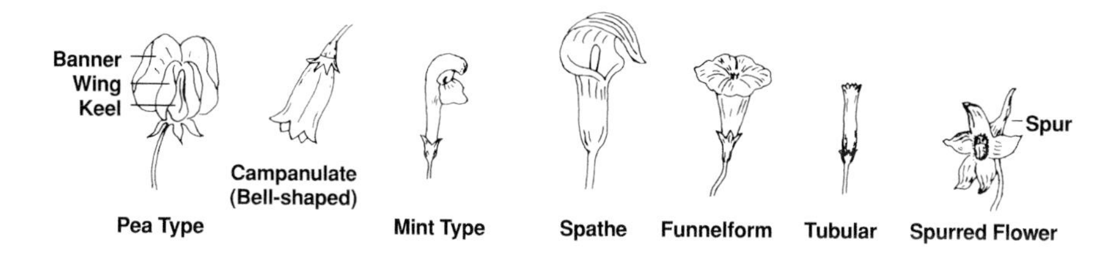

INFLORESCENCES

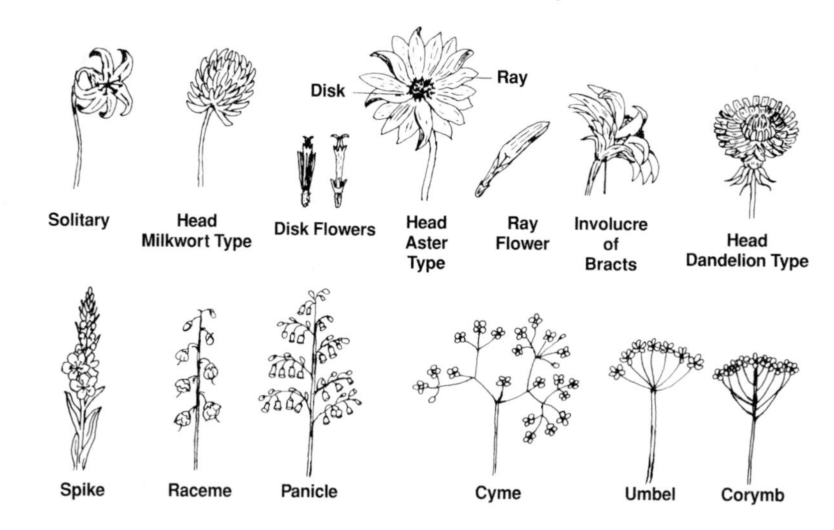

Flower parts, flower types, and inflorescences.

LEAF ARRANGEMENTS

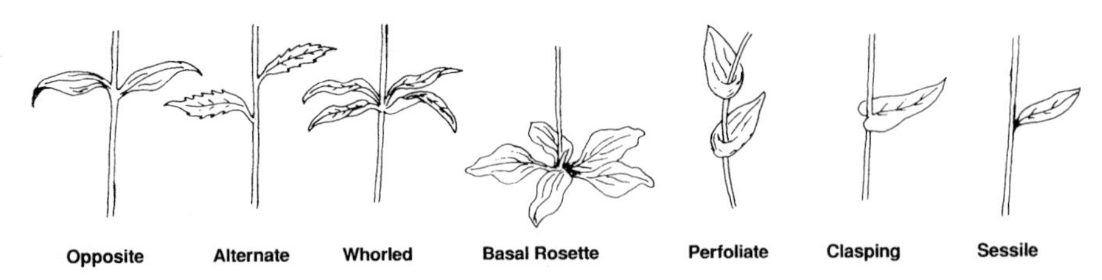

LEAF DIVISIONS

SHAPES

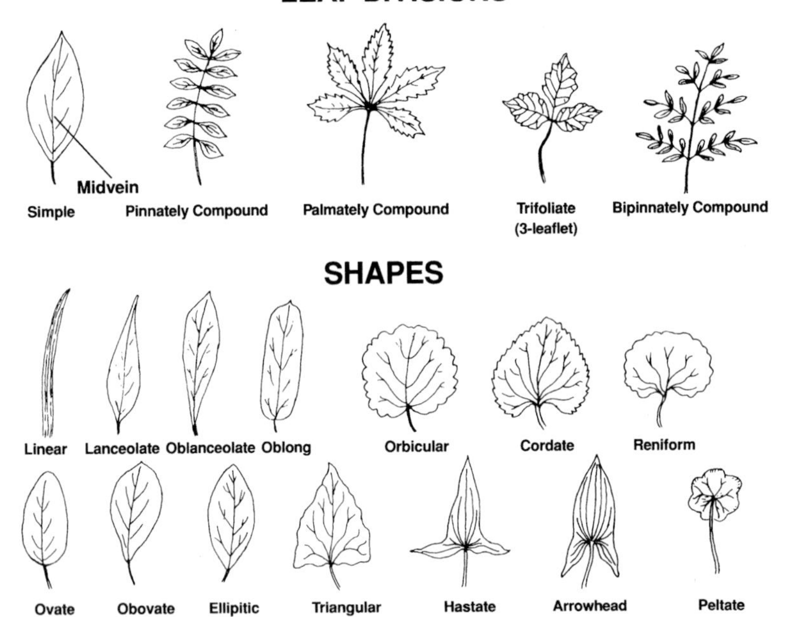

MARGINS

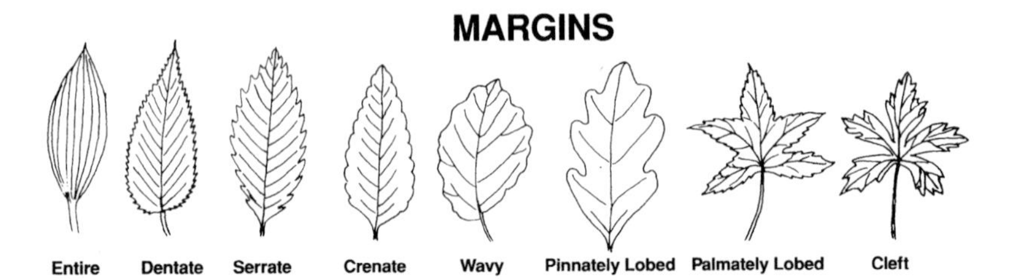

Leaf arrangements, leaf divisions, shapes, and margins.

Used by permission of the University of Alabama Press

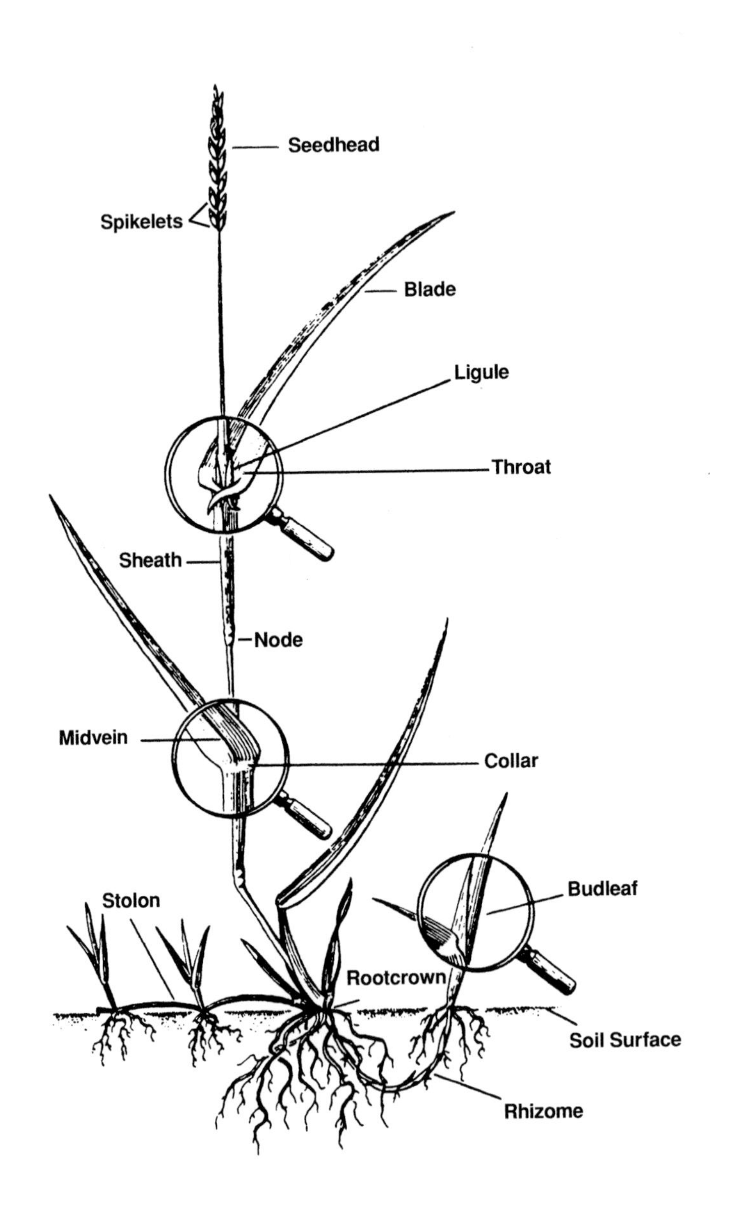

Parts of a grass plant.

Used by permission of The Scotts Company

Miller, James H., Chambliss, Erwin B., and Loewenstein, Nancy J. 2010. A field guide for the identification of invasive plants in southern forests Gen. Tech. Rep. SRS–119. Asheville, NC: U.S. Department of Agriculture, Forest Service, Southern Research Station. 126 p.

Invasions of nonnative plants into forests of the Southern United States continue to go unchecked and only partially unmonitored. These infestations increasingly erode forest productivity, hindering forest use and management activities, and degrading diversity and wildlife habitat. Often called nonnative, exotic, nonindigenous, alien, or noxious weeds, they occur as trees, shrubs, vines, grasses, ferns, and forbs. This book provides information on accurate identification of the 56 nonnative plants and groups that are currently invading the forests of the 13 Southern States. It lists other nonnative plants of growing concern. Recommendations for prevention and control of these species are provided in a companion booklet, "A Management Guide for Invasive Plants of Southern Forests," published by the Southern Research Station as a General Technical Report. Basic strategies for managing invasions on a specific site include maintaining forest vigor with minimal disturbance, constant surveillance and treatment of new unwanted arrivals, and finally, rehabilitation following eradication.

Keywords: Alien plants, exotic weeds, forest noxious plants, invasive exotic plants, invasive nonindigenous plants.